Life Support Systems for Humans in Space

Erik Seedhouse

Life Support Systems for Humans in Space

Erik Seedhouse
Embry Riddle Aeronautical University
Daytona Beach, FL
USA

ISBN 978-3-030-52858-4   ISBN 978-3-030-52859-1   (eBook)
https://doi.org/10.1007/978-3-030-52859-1

This Springer imprint is published by the registered company Springer Nature Switzerland AG
The registered company address is: Gewerbestrasse 11, 6330 Cham, Switzerland

Preface

» The temperature was not strikingly low as temperatures go down here, but the terrific winds penetrate the flimsy fabric of our fragile tents and create so much draught that it is impossible to keep warm within. At supper last night our drinking-water froze over in the tin in the tent before we could drink it. It is curious how thirsty we all are.

— Ernest Shackleton, *South*

I teach a number of life support systems classes at the undergraduate and postgraduate level. After every course, students are encouraged to write course evaluations, and one of the most repeated comments in these evaluations is that there is no dedicated textbook on the subject of spaceflight life support systems. Peter Eckart's excellent *Spaceflight Life Support and Biospherics* addresses some of the material covered in the courses I teach, but the book was published way back in 1996, and a lot has changed since then.

So, I decided to write the book that you're holding now. This book is intended to support the myriad courses I teach on the subject of spaceflight life support in addition to other courses I teach that include the subject of life support. Hopefully, this book will also be helpful as a reference guide for courses taught at other universities and colleges around the world. The book is structured as a textbook and follows a step-by-step approach in addressing the core topics of spaceflight life support. In addition to serving as a textbook, this publication is also intended to be a source of information on the subject of spaceflight life support and the development of technologies that enable astronauts to spend months in space.

This book cannot encompass all the topics of spaceflight life support. For that, there is a major reference work – the *Handbook of Life Support Systems for Spacecraft and Extraterrestrial Habitats* – for which I am a Co-Editor-in-Chief. This major reference work is published by Springer and is nearing completion at the time of writing.

As always, in writing any book, I have been fortunate to have had reviewers (in this case via Praxis, for whom I have written many books) who made helpful comments concerning the content of this publication. I am also grateful to Maury Solomon and Hannah Kaufman and their team at Springer for guiding this book through the publication process. Thank-you to Ms. ArulRonika Pathinathan, Project Manager at Springer/SPi Content Solutions, and her team in producing this textbook and thank-you also to all those who gave permission to use many of the images in this book.

Erik Seedhouse, PhD
Flagler Beach, FL, USA
August 2020

Contents

About the Author

Erik Seedhouse
is a Professor of Spaceflight Operations and Human Factors Aviation Safety at Embry-Riddle Aeronautical University. He has extensive practical and theoretical experience in many of the subjects in this book. After completing his first degree, he joined the 2nd Battalion, the Parachute Regiment. During his time in the "Paras," Erik spent 6 months in Belize, where he was trained in the art of jungle warfare. Later, he spent several months learning the intricacies of desert warfare in Cyprus. He made more than 30 jumps from a Hercules C130 aircraft, performed more than 200 helicopter abseils, and fired more light anti-tank weapons than he cares to remember!

Upon returning to academia, the Erik embarked upon a Master's degree which he supported by winning prize money in 100 km running races. After placing third in the World 100 km Championships in 1992, Erik turned to ultra-distance triathlon, winning the World Endurance Triathlon Championships in 1995 and 1996. For good measure he won the World Double Ironman Championships in 1995 and the infamous Decatriathlon, an event requiring competitors to swim 38 km, cycle 1800 km, and run 422 km. Nonstop!

In 1996, Erik pursued his PhD at the German Space Agency's Institute for Space Medicine. While studying, he found time to win Ultraman Hawai'i and the European Ultraman Championships as well as completing Race Across America. Due to his success as the world's leading ultra-distance triathlete, Erik was featured in dozens of magazine and television interviews. In 1997 GQ magazine named him the "Fittest Man in the World."

In 1999 Erik took a research job at Simon Fraser University. In 2005, he worked as an astronaut training consultant for Bigelow Aerospace. Between 2008 and 2013, Erik served as Director of Canada's manned centrifuge and hypobaric operations. In 2009, he was one of the final 30 candidates in the Canadian Space Agency's Astronaut Recruitment Campaign. Erik has a dream job as a professor at Embry-Riddle Aeronautical University in Daytona Beach, Florida. He holds a pilot license and in his spare time works as an astronaut instructor for Project PoSSUM, occasional film consultant to Hollywood, a professional speaker, triathlon coach, and author. He also serves as a consultant to myriad television productions, and in the summer he usually ventures into the mountains—he has reached the summit of Kilimanjaro, Aconcagua, Elbrus, Rainier, Island Peak, and several others. This textbook is his 30th and final publication (one of his bucket list items was to write a bookshelf of books and Erik's bookshelf holds 30 books, so objective achieved!). When not enjoying the sun and rocket launches on Florida's Space Coast with his wife, Alice, he divides his time between Waikoloa, Hawai'i and his cottage in Sandefjord, Norway.

Life Support System Basics

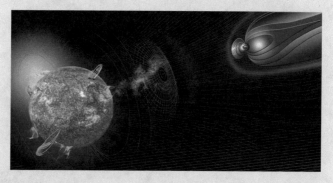

Credit: NASA

Contents

© Springer Nature Switzerland AG 2020
E. Seedhouse, *Life Support Systems for Humans in Space*,
https://doi.org/10.1007/978-3-030-52859-1_1

🔄 Learning Objectives

After completing this chapter, you should be able to:

- Describe and distinguish between the layers of the atmosphere and provide salient characteristics of each layer
- Describe key characteristics of the lithosphere, hydrosphere, and biosphere
- Explain what is meant by the term *biomes*
- Explain the significance of the carbon and phosphorus cycles as they pertain to life on Earth
- Describe key characteristics of the space environment
- Explain the difference between SPEs and GCRs
- Explain what is meant by RBE
- Describe the rationale for tissue weighting when assessing radiation exposure
- Explain the effects of microgravity on human physiology
- Explain the significance of the Armstrong Line
- List five key distinguishing characteristics of the lunar and Martian environments

Introduction

Any book that tackles the subject of life support has to start with a summary of the basic features of the terrestrial environment here on Earth. So, the first part of this chapter does exactly that (**◻** Fig. 1.1). We'll begin with the atmosphere.

◻ Fig. 1.1 Crew Earth Observation image ISS008-E-13304 taken on 28 January 2004. This is one of the top ten most popular images taken by astronauts from the International Space Station (ISS). The image features Mt. Everest (8850 meters) and Makalu (8462 meters) and an oblique view of the Himalayas looking south from over the Tibetan Plateau. Credit: NASA

1

The Atmosphere

One way to think of the Earth is to think of it being comprised of several layers. Let us start with the atmosphere (◘ Fig. 1.2). As you can see in ◘ Fig. 1.2, the lowest layer of the atmosphere, the *troposphere*, extends to only 17 kilometers altitude, but it is this layer that contains most of our life-giving oxygen. As you climb higher in the troposphere, temperature and air pressure fall.

The next layer is the *stratosphere*, which stretches to 48 kilometers above sea level. The lower portion of the stratosphere contains ozone, which is important for filtering out much of the Sun's harmful ultraviolet radiation. Unlike the troposphere, the stratosphere *increases* in temperature as you climb higher. This rising temperature trend means that air in the stratosphere is not as turbulent as the troposphere, which is one of the reasons commercial aircraft fly in this layer of the atmosphere.

Continue climbing and you reach the *mesosphere*, which extends to 85 kilometers altitude. Unlike the stratosphere, this layer gets colder the higher you rise. In fact, the coldest temperatures (about −90 °C) in Earth's atmosphere are found near the upper reaches of this layer.

◘ **Fig. 1.2** The divisions of the atmosphere. Credit: NASA

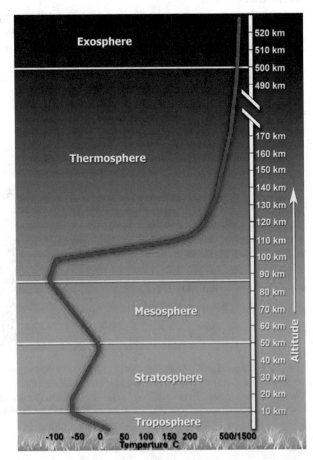

◘ Fig. 1.3 The Northern lights. Credit: NASA

Next up is the *thermosphere*, which is a layer that absorbs high energy X-rays and ultraviolet radiation. Since the Sun exerts such an influence on this layer, the upper limit of this layer may vary between 500 and 1000 kilometers (the ISS orbits at an altitude of 400 kilometers). Temperatures can vary between 500 °C and 2000 °C in this layer. It also happens to be the layer in which the Northern and Southern lights occur (◘ Fig. 1.3).

We're still not at the top of the atmosphere. Next is the final layer: the *exosphere*. This layer is the final frontier of Earth's atmosphere, and as you can imagine, the air in this layer is next to nonexistent. It's thin. *Really* thin. Its upper limit stretches out to somewhere between 100,000 and 190,000 kilometers above the surface of the Earth.

The Lithosphere, Hydrosphere, and Biosphere

So that's the atmosphere, but what about the area near the surface of the Earth? Well, this area is best thought of as a series of interconnected spheres. Four of them.

The first of these, the *lithosphere*, comprises the rocks, the planet's mantle, and crust. Mount Everest, the beaches of the Maldives, and Mauna Kea (the highest mountain on Earth from base to summit incidentally) are all components of the lithosphere.

1

The *hydrosphere* comprises all the water on or near the Earth's surface. Ninety-seven percent of all the water on our planet is found in the oceans, with the remaining fresh water found in lakes and the polar regions. But water doesn't exist in a static environment. Instead, it changes as it moves through the hydrological cycle (◙ Fig. 1.4).

The *biosphere* comprises all living organisms, most of which are found in a zone that stretches from 3 meters below ground to about 30 meters above ground. In the oceans, the majority of aquatic life can be found in a zone that stretches from the surface to about 200 meters below. Of course, some creatures can happily survive outside these zones. For example, some species of fish have been found as deep as 8 kilometers, and then there are the *extremophiles* that can be found in acidic environments such as Mono Lake.

Within the biosphere are *biomes*, which are regions of the Earth that have similar climate, animals, and plants. The biomes are divided into *terrestrial biomes* (desert, forest, grassland, and tundra) and *aquatic biomes* (freshwater and aquatic). Why is understanding all this is important in the context of spaceflight life support? Well, as we shall see, sustaining life on Earth is achieved thanks to deviously complex and interrelated processes. It is only by having an appreciation of these processes that we can even begin to fathom just how difficult it is to replicate them in the confined environment of a spacecraft. So let's continue.

Warming the biosphere is the Sun, which also supplies the energy for photosynthesis (more about this in ▶ Chap. 8), provides the power for the cycling of matter, and drives the weather systems. Most of the sunlight that reaches the troposphere is visible light, and about a third of the solar energy that reaches this layer of the atmosphere is reflected back to space by clouds and the Earth's surface. The ability of surfaces to reflect radiation is referred to as *albedo*, and it is determined by color and texture. For example, ice and snow have a high albedo, whereas forests have a low albedo.

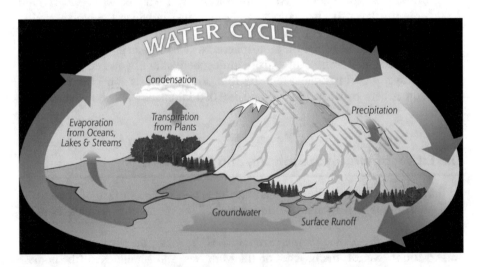

◙ **Fig. 1.4** The water cycle. Credit: NASA

Matter Cycling in the Biosphere

The next characteristic we must consider is the process by which organisms grow inside the biosphere: matter cycling. For organisms to live, grow, and reproduce, nutrients are required. When nutrients are required in large amounts, these nutrients are termed *macronutrients*, whereas trace elements required by organisms are termed *micronutrients*. An example of a micronutrient is iron.

The Oxygen Cycle

Regardless of which nutrient element is required, these elements continuously cycle from the air, water, and soil to organisms and back to the air, water, and soil in a process termed a *nutrient cycle*. These biogeochemical cycles, which are driven directly by solar energy and gravity, include specialized cycles such as the oxygen, nitrogen, and hydrologic cycles. For example, the *oxygen cycle* (□ Fig. 1.5) describes the circulation of oxygen in its various forms in the biosphere. In this cycle, plants and animals use oxygen to respire. In this process of respiration, oxygen is returned to the atmosphere as carbon dioxide, which is metabolized by algae (this is covered in more detail in ► Chap. 8) and plants, and then converted into carbohydrates during photosynthesis, with oxygen being the by-product. In our biosphere, the oceans are the biggest generators of oxygen.

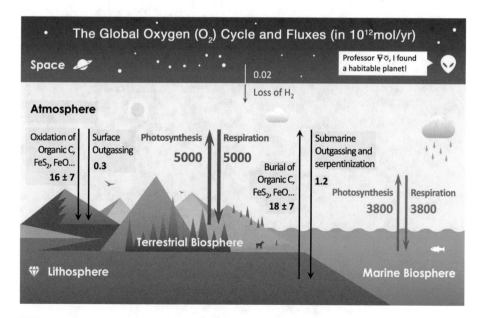

□ **Fig. 1.5** The oxygen cycle. Credit: NASA

The Nitrogen Cycle

1

As we know, oxygen is essential to sustain life, but nitrogen, which is the most plentiful element in Earth's atmosphere, is also essential to human survival. Yet even though we're surrounded by nitrogen, animals and plants cannot utilize nitrogen in its free form, because they don't have the enzymes required to convert the nitrogen to a form that can be used. But through *bacteria*, free nitrogen can be combined chemically with other elements to form usable (i.e., more reactive) compounds such as ammonia and nitrites. This process, which is referred to as *nitrogen fixation*, forms a key part of the nitrogen cycle (◘ Fig. 1.6). Most nitrogen fixation is achieved by certain types of bacteria and algae. The products of nitrogen fixation include compounds that can be used by the tissues of algae and plants. So, animals that eat these algae and plants metabolize these compounds; the by-products, such as urea, are excreted; and then they are ultimately converted to ammonia and eventually nitrates and nitrites. Finally, the nitrates and nitrites undergo a conversion to atmospheric nitrogen, thanks to the action of denitrifying bacteria, and the whole cycle starts again.

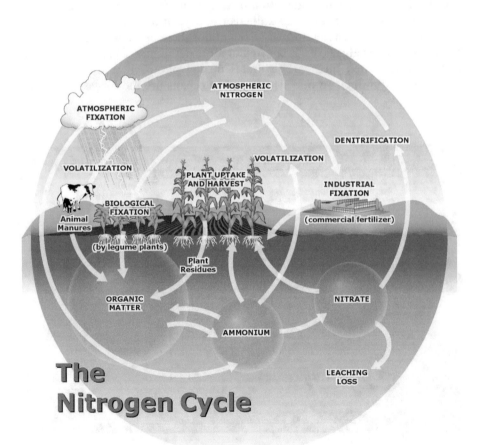

◘ **Fig. 1.6**   The nitrogen cycle. Credit: NASA

The Carbon Cycle

Another key cycle we must mention is the carbon cycle. As we know, carbon is the basic building block of carbohydrates, fats, and proteins, but it is also essential to nucleic acids such as DNA and RNA. The carbon cycle is driven by carbon dioxide, which is found in seawater and rock and which comprises about 0.035 percent of the lower level of the atmosphere. What follows is a brief description of how this cycle works.

First, carbon is attached to oxygen in the atmosphere and becomes carbon dioxide. Then, thanks to photosynthesis, carbon dioxide is extracted from the atmosphere and enters the food chain. Carbon's first step in the food chain is in plants, which are eaten by animals. When animals and plants die, the decay brings the carbon into the ground. There, they become fossil fuels that are burned, resulting in carbon being released into the atmosphere. Another way by which carbon is released into the atmosphere is by you breathing. Every time you exhale, you release carbon dioxide – 1 kilogram of the stuff every day (incidentally, getting rid of carbon dioxide is one of the toughest tasks of a spacecraft life support system). The oceans also play a role in carbon dioxide regulation. Some carbon dioxide remains in the seawater, but some is removed by marine ecosystems that absorb carbon dioxide and form carbonate compounds such as calcium carbonate to build shells.

The Phosphorus Cycle

The final cycle we'll discuss is the phosphorus cycle. Why phosphorus? Well, this nutrient is not only essential for plants and animals, but it also is a part of DNA. Phosphorus is released by the gradual breakdown of phosphate rock deposits and is dissolved in soil water before being used by plants. Animals get their phosphorus intake by eating plants, and the cycle comes full circle when the animals die and the decay products return phosphorus to the soil.

Space Environment

No discussion of the space environment can begin before establishing where space actually starts. If you asked a USAF pilot back in the 1960s, he (because there were no female pilots back then) would have told you that 50 miles was the magic altitude. After all, if pilots reached this altitude, which they often did when flying the X-15 (◘ Fig. 1.7), the air force awarded them astronaut wings.

Nowadays, the altitude at which space starts is set at 100 kilometers, which happens to be 62.21 miles or 328,000 feet in old money. Some of you may remember way back when in 2004 there was a competition called the X Prize. It called for a privately developed spacecraft to reach 100 kilometers and repeat the feat within 2 weeks. A spacecraft dubbed SpaceShipOne (SS1) cobbled together by Burt Rutan and his Scaled Composites team won the competition ($10 million, although the cost of achieving the win was $25 million, but that's another story). SS1 morphed

1

◘ **Fig. 1.7** X-15. Credit: USAF

into SpaceShipTwo (SS2) under the new ownership of Richard Branson's Virgin Galactic, and for more than a decade and a half, this company promised to take space tourists to space. This is a promise probably destined never to be fulfilled. Why? Because in the small print (always read the small print – *always!*) on the ticket, Virgin Galactic promises to take its passengers to an altitude of at least 50 miles. Close, but no cigar. Anyway, the point is that the altitude of space is somewhat arbitrary, so let's move on.

Radiation

Once you get into space, there are a number of things that grab your attention. One of these is the Sun. It's basically a big yellow ball in the blackness of space fueled by nuclear fusion. The Sun creates all sorts of problems for astronauts and life support engineers, one of which is radiation. As far as Suns go, our Sun is pretty ordinary. Just one small yellow star out of billions in our galaxy. Powered by aforesaid nuclear fusion, our Sun fuses about 600 million tons of hydrogen *every second*. Two by-products of this fusion process include electromagnetic (EM) radiation and charged particles, and it is these by-products that comprise the radiation that occurs in our Solar System. That light and heat you feel on your face on a bright summer day is EM, whereas the other fusion by-products are those charged particles such as protons and electron. Protons are particles that have a positive charge, whereas electrons have a negative charge.

During fusion, the Sun's interior generates heat, and this heat is so intense that a fourth state of matter is created. We're all familiar with the three states of matter (solid, liquid, and gas), but if heating of these states continues, molecules eventu-

■ **Fig. 1.8** A solar particle event. Credit: NASA

ally begin to break down, and eventually the atoms will form a plasma. Inside the Sun, this is exactly what happens, but the charged particles that comprise this plasma are disturbed by the intense magnetic field, and these charged particles shoot away from the Sun at up to 700 kilometers per second. It is this stream of charged particles that is termed the solar wind. Every once in a while, areas of the Sun's surface become more active than usual, causing the surface to spew out bursts of these charged particles. These bursts are termed *solar particle events* (SPEs) (■ Fig. 1.8). These SPEs may last for a few hours or a few days. Either way, they are violent events that may occasionally reach Earth's orbit and they can potentially kill astronauts.

The other type of radiation is *galactic cosmic radiation* (GCR). GCR originates outside the solar system and comprises the remnants of exploding stars. This radiation is comprised of extremely energetic particles such as hydrogen, iron, and helium. The nuclei of these particles are fully ionized, which means all the electrons have been stripped from the atoms. In turn, this means that these particles will interact with magnetic fields. But what makes GCRs so devastating is that these charged particles are zipping along at close to light speed and therefore have incredible energy. GCR (■ Fig. 1.9) is the most dangerous category of radiation that astronauts and life support engineers have to worry about. How dangerous? Well, of the radiation that astronauts are exposed to, 99 percent is generated by the Sun, and 1 percent is GCR, but 99 percent of the damage is caused by that 1 percent. And because this radiation is so damaging, it makes sense to track it, which is why ISS modules have dosimeters to do this job.

Tissue Weighting

Dosimeters are used to measure the dose of radiation astronauts are exposed to. A dose is simply the amount of energy deposited in the various body tissues by radiation. But not all tissues are affected by radiation equally, which is why the concept of *relative biological effectiveness* (RBE) was developed. Basically, in applying this concept, each type of radiation has a RBE: the higher the number, the more damaging the radiation. For example, X-rays have an RBE of 1, whereas alpha particles may have an RBE between 10 and 20, making these particles especially damaging.

1

But simply measuring the dose does not tell life support specialists everything they need to know about how damaging radiation is to the body [1–3]. To do that, a *quality factor* is applied. This quality factor is Q, more commonly known as a weighting factor (W_R). W_R is a function of linear energy transfer (LET), which is the amount of energy an ionizing particle of radiation deposits in a material by distance. In other words, LET is a way of describing the action of radiation as it passes through matter. How is radiation measured? Well, it depends on the context. *Grays* are used to measure the energy absorbed per unit of mass, while *roentgens* are used to measure exposure to X-rays. *Sieverts* (Sv) are used to measure what is termed the equivalent dose (H), and they are also used to measure the effective dose (E). You may have come across rads and rems, but these are non-SI units, so we'll steer clear of these.

So, let's get back to H and E. First, we'll talk about H. This simply describes the absorbed dose required to produce a biological effect, and the magnitude of the effect will be determined by different types of radiation. To make the accuracy of the effect of radiation as precise as possible, H is multiplied by W_R. By doing this, the RBE is factored in, which results in a reasonably good estimation of the effects of radiation on tissues such as organs. But while H provides useful data about radiation dose, it is E that is more accurate when it comes to specifying exposure limits, and this data is important to ensure astronauts are exposed to radiation within acceptable levels. The specificity of E lies in the application of specific weighting factors (◘ Table 1.1) for each tissue (expressed W_T), which allows for an accurate whole-body dose to be calculated (see Appendices Ia and Ib).

Another radiation source we must mention is the Van Allen belts. These doughnut-shaped belts comprise highly energetic charged particles that surround the Earth. Discovered by James Van Allen in 1958, these radiation belts have been the focus of intense study over the decades. We know, for example, that the inner belt is the primary source of radiation for spacecraft that orbit above 500 kilometers altitude. We also know that the belts expand with increased solar activity and that the

◘ Table 1.1 Tissue weighting factors for organs

Organs	Tissue weighting factor
Gonads	0.08
Red bone marrow	0.12
Colon	0.12
Lung	0.12
Stomach	0.12
Breasts	0.12
Bladder	0.04
Liver	0.04
Esophagus	0.04
Thyroid	0.04
Skin	0.01
Brain	0.01
Remainder of the body	0.12

outer belt comprises electron and protons. What isn't so well understood is what happens when solar particles strike the belts during a geomagnetic storm. This is important because solar storms can short-circuit spacecraft and communications can be disrupted. We also don't understand exactly how exposure to Van Allen radiation affects astronauts. That isn't surprising because only the moon-bound Apollo astronauts have transited through the belts (the trajectory was designed to pass through the thinnest parts of the belts).

Gravity

Whenever we see images of astronauts floating (◘ Fig. 1.10) around the ISS, we often hear the term *zero gravity*, but this is a misleading misnomer. A more accurate term is *microgravity*, which is a condition in which astronauts *appear* to be in a state of zero gravity. Why is this so?

As we know, gravity causes objects to pull other objects toward it. Gravity is what keeps the Moon in orbit around the Earth. But gravity becomes weaker with distance. The ISS orbits the Earth at an altitude of around 400 kilometers. At this altitude, Earth's gravity is approximately 90 percent of what it is on the surface. Let us put this another way. If an astronaut who weighs 70 kilograms on the Earth's surface could be magically transported to the ISS, that astronaut would weigh 63 kilograms. Now you may be wondering how astronauts float inside the ISS when 90 percent of Earth's gravity is still affecting the space station [4]. Well,

1

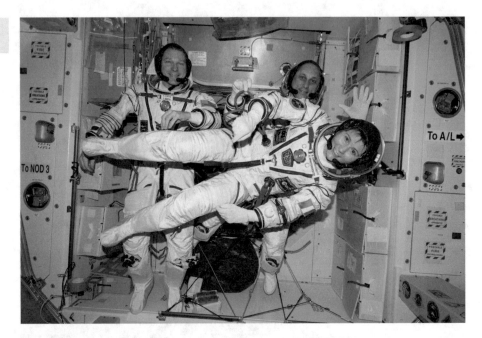

☐ Fig. 1.10   ISS crewmembers floating in the Unity module. Credit: NASA

that's because the crew are in free fall. In a vacuum (the ISS is not technically in a vacuum, but we'll get to that shortly), gravity causes objects – including astronauts – to fall at the same rate. But since they are in space, astronauts are not falling toward Earth, but *around* it. The reason the ISS doesn't fall to Earth is because it is moving quickly. Really quickly. Around 7.8 kilometers per second! This just happens to be the right speed to match the curve of the Earth.

So far so good, right? But this microgravity is anything but good for the astronauts. As we shall see later, microgravity causes havoc on human physiology. In just 6 months on board the ISS, astronauts lose between 20 and 25 percent of muscle mass and more than 6 percent of their bone mass density, and they suffer vision impairment, intracranial pressure, headaches, fluid shifts, and the list goes on and on.

Space Debris

Space junk isn't a topic that immediately comes to mind when talking about life support systems, but protecting astronauts from debris is a serious life support issue. As for most things space-related, NASA has an acronym for space debris, MMOD, which stands for micrometeoroids and orbital debris [5, 6]. You may think the chances of an astronaut getting hit by a piece of space junk are pretty remote, but the truth is that the *low Earth orbit* (LEO) environment is clogged with space debris. With every space mission, bits and pieces are left in space. Right now there is about 2200 tons of debris floating (☐ Fig. 1.11) around LEO, and since this junk is moving at speeds of 7 kilometers per second, getting hit by even a small flake

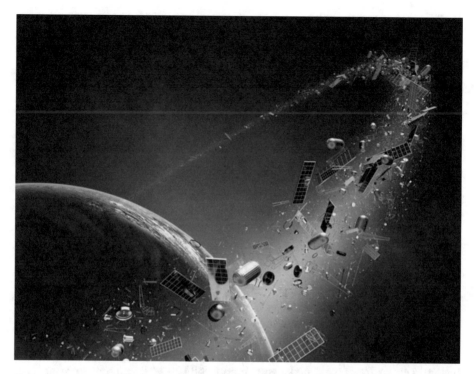

□ Fig. 1.11 Space debris is a serious and evolving hazard. Credit: NASA

□ Fig. 1.12 Impact crater caused by a flake of paint hitting the windshield of the Shuttle during STS-7. Credit: NASA

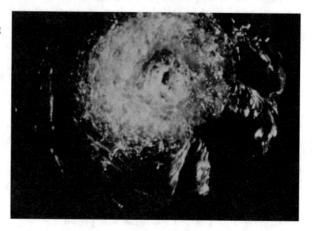

of paint can be a big deal because even such a tiny chunk of material can pack a punch greater than a bullet. Not convinced? Take a close look at □ Fig. 1.12. That crater in the Space Shuttle window that you can see was caused by a flake of paint. Just *one* flake. Of paint! The crater, which measured 0.2 mm in depth and 4 mm in width, didn't penetrate the Shuttle, but it brought to light just how common collisions with spacecraft are. Incidentally, the repair job for the incident ran to $50,000.

1

Crews on board the ISS have also had to deal with their fair share of space junk encounters. In March 2012, newspapers ran the banner *Space Junk Forces Station Astronauts to Take Shelter in Lifeboats* when a piece of old Russian satellite was spotted on a trajectory that appeared uncomfortably close to call. So close that mission control ordered the crew to seek shelter in the two Soyuz spacecraft, just in case. The errant piece of space junk eventually zipped by 11 kilometers from the station, meaning astronauts could breathe a sigh of relief. It was the third time in 12 years that crews had had to take shelter from a close call with space debris. NASA and the station partners usually have the astronauts position the station in an avoidance maneuver when space junk is expected to pass, but the March 2012 incident (which turned out to be a debris remnant of the Russian Cosmos 2251 satellite) caught the agency, and also the US military's Space Surveillance Network (SSN), napping. After all, it is the SSN's job to track the thousands of objects in LEO.

Vacuum

Space is often portrayed by those in Hollywood as a cold and inhospitable place at the best of times, and the fact that there is no air is one of the features that has given rise to all sorts of fallacies portrayed in science fiction films such as *Prometheus*, *Total Recall*, and *Outland*. In these films, inadvertent exposure to a vacuum will instantly either freeze you, make your blood boil (*Total Recall*), make your head explode (*Outland*), or all three! So let's put some of those fallacies to rest.

First: the cold [4, 7, 8]. This is all down to thermodynamics. Temperature is a function of heat energy in a given amount of matter and space. But space has no matter. That's why it's called space! And according to thermodynamics, heat transfer cannot happen in space because conduction and convection cannot occur without matter.

So if you happen to step out of an airlock without a spacesuit (not advisable!), what would happen? Well, if the spacecraft you just stepped out of was in sunlight, you would feel warm, whereas if you were shielded from the sun, you would feel cold. If you happened to be somewhere in deep space where the temperature is a chilly −270 °C, you still wouldn't freeze. Not instantly at any rate, because heat transfer can't happen as quickly as radiation alone.

So, that's the first part of the "stepping into a vacuum" experience. Now, what else can our unfortunate suitless astronaut expect after their sudden decompression to a vacuum (incidentally, this is termed *ebullism* in life support parlance)? First, a reminder about some laws of physics.

We'll begin with Henry's Law, which states that "at a constant temperature, the amount of gas that dissolves in a given volume of liquid is directly proportional to the partial pressure of the gas in equilibrium with that liquid." Or, in other words, "the solubility of gas in a liquid is directly proportional to the partial pressure of the gas above said liquid." Why is this law important in the context of jumping suitless into space? The vapor pressure of water at body temperature is about 6 percent of atmospheric pressure or about 47 mmHg. Below this pressure, which is a pressure you would find at around 19,000 meters (63,000 feet – a physiological boundary known as the Armstrong Line), bodily fluids begin to boil away.

The Armstrong Line has *nothing to do with Neil Armstrong*. This physiological boundary was named in honor of Harry George Armstrong (◘ Fig. 1.13), who was a major general in the United States Air Force. Armstrong is widely recognized as one of the leading pioneers of aviation medicine, which is why the *Armstrong Line* or *Armstrong Limit*, which is the altitude above which water boils at human body temperature, was named after him. During his career, Armstrong served as director of the United States Aeromedical Laboratory, where he applied his aviation knowledge to protecting aircrew from hypoxia at high altitude.

◘ **Fig. 1.13** Harry Armstrong. Credit: NASA

1

This last fact is one that can have dire consequences in the event of a pressure suit breach. I was lucky enough to wear a pressure suit during an Armstrong Line test many years ago. After donning said suit, I was led into a hypobaric chamber, and a glass of water was placed at eye level opposite my seat. Then I was sealed into the suit, the entrance hatch to the chamber was locked and the air was sucked out. As the altimeter began to approach 60,000 feet, the water in the glass began to boil. By the time the chamber had leveled off at 70,000 feet, the water had practically boiled away, which is exactly what would have happened to my blood if my suit had suffered a breach. After a few minutes contemplating my mortality and performing some confidence tests in the suit, air was pumped back into the chamber to return it to ground level. A lot of fun!

But let's get back to our suitless astronaut. Deep inside this hapless astronaut's body, water would begin to turn into gas and become water vapor. This process, which would happen rapidly in the lungs and beneath the skin, would also lead to bubbles of water vapor forming in venous blood. This latter process would be so profound that the circulation would effectively be vapor-locked. But the worst would be yet to come. The precipitous drop in pressure would cause the air in the lungs to expand rapidly. The best course of action for our astronaut would be to keep their mouth open to ensure that the air rushed outward. If our astronaut didn't do this, then lung rupture would be the consequence. Not that it really would matter, because with no circulation (it's vapor-locked remember!) and no oxygen, our astronaut would only be conscious for about 12 seconds. A similar fate befell cosmonauts Dobrovolski, Patsayev, and Volkov (☐ Fig. 1.14) who, in

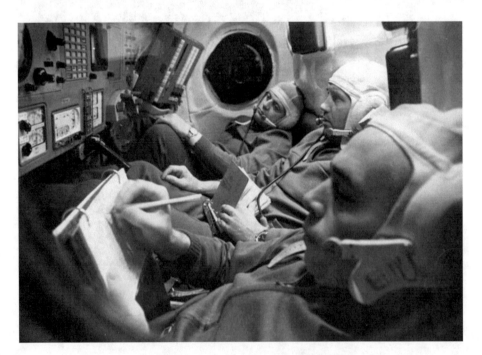

☐ **Fig. 1.14**   Cosmonauts Dobrovolski, Patsayev, and Volkov. Credit: Roscosmos

1971, died as a result of high altitude decompression when their Soyuz spacecraft malfunctioned.

For those who regard themselves as science fiction aficionados, you will no doubt recall Dave Bowman's predicament in *2001: A Space Odyssey*. Bowman, faced with the problem of entering a spacecraft after being locked out by the malevolent HAL 9000 computer, ejects himself through space into the unpressurized airlock of the spacecraft. Once inside the airlock, Bowman activates the handle and repressurizes the airlock. Total time for all events according to my stopwatch: about 12 seconds. So, if Bowman had had the presence of mind to exhale during decompression and to keep his mouth open during his exertions, it is probable he could have accomplished the feat and lived, since the scenario following the decompression ends with a pressurized finale. Incidentally, the scene is so accurate that I show it to students as a demonstration that explosive decompression doesn't have to be messy.

Planetary Environments

Now that we have discussed a few characteristics of our planetary environment and select features of the space environment, it is time to turn our attention to some local planetary environments. As you may know, there are NASA-funded plans afoot to return to the Moon (◘ Fig. 1.15) by the mid-2020s and commercial (SpaceX) plans to visit Mars before the end of the 2020s. Manned missions further afield than the Moon or Mars before the 2050s are extremely unlikely, so we'll restrict our focus accordingly.

◘ Fig. 1.15   A future lunar base. Credit: NASA

1

The Moon

To begin with, the Moon is a lot smaller than the Earth, and its gravity is only one-sixth of Earth's gravity. Due to its very low escape velocity, the Moon can't maintain a significant atmosphere, which means the surface is exposed directly to a vacuum. The lack of an atmosphere also means surface temperatures can be extreme. When in direct sunlight, the temperature can reach 107 °C, and when the Sun sets, the temperature plummets to −153 °C (◘ Table 1.2). That's a major headache for life support engineers. An even bigger headache is the lunar regolith, a by-product of heavy meteoroid bombardment over billions of years. This regolith caused havoc for the Apollo astronauts, who suffered pulmonary problems and significant equipment wear and tear. And then there is the danger posed by those meteoroids.

As if that wasn't enough, radiation presents yet another hazard. First there is the solar wind, which is plasma traveling at around 400 kilometers per second. That plasma is comprised of charged particles such as electrons and protons. Radiation levels can also be affected by solar flares, which may possess even greater energies than the solar wind. And then there is GCR which, as we have already learned, are very high energy particles comprising protons, electrons, positrons, and gamma rays [9–11].

Since the Moon has neither a magnetic field nor an atmosphere, all those computer-generated images of future astronauts bouncing around on the surface of the Moon are fanciful in the extreme. The reality will be astronauts living in bunkers (take a close look at that habitat being constructed in ◘ Fig. 1.15) shielded not only from the deadly onslaught of killer radiation but also from the extreme temperature range and bombardment from meteoroids.

So what about surface operations? Here, moon suits have many disadvantages. Firstly, they take time to don and doff. Secondly, they have extremely limited duty cycles when exposed to the corrosive regolith. Thirdly, consumables, recycling sys-

◘ **Table 1.2**   Select properties of the Moon (with Earth as comparison)

	Moon	**Earth**
Mean radius	1,737 km	6,378 km
Surface area	37,900,000 km^2	510,000,000 km^2
Mass	0.0735 × 10^{24} kg	5.976 × 10^{24} kg
Mean density	3.34 g/cm^3	5.52 g/cm^3
Mean surface gravity	162 cm/sec^2	980 cm/sec^2
Escape velocity	2.38 km/sec	11.2 km/sec
Mean surface temperature	day, 107 °C night, −153 °C	15 °C
Temperature extremes	123 °C to −233 °C	56.7 °C to −89.2 °C
Surface pressure	3 × 10^{-15} bar	1 bar

tems, and operator fatigue are extremely limiting in terms of how long a surface EVA can last. Fourthly, the length of time to repair and refurbish suits means these suits can only be used in extreme circumstances. Fifthly, no agency wants their astronauts unnecessarily exposed to the killer trio of radiation exposure, vacuum, and meteoroid risks. So, the reality will be astronauts tele-operating rovers and other surface assets from the relative safety of an underground 3-D printed bunker [12].

Mars

Mars is a destination that has been on the drawing board for decades. NASA has published design reference missions, and there have been myriad commercial ventures, including but not limited to *Inspiration Mars*, *Mars One*, and, more recently, Elon Musk's Mars Venture. So why have no humans landed on Mars yet?

From a life support perspective, Mars is as much a challenge as the Moon, although it does have some advantages, one of which is an atmosphere (■ Table 1.3). The Martian atmosphere is about 100 times thinner than Earth's (about equal to being 130,000 feet above the surface of the Earth), and it is 95 percent carbon dioxide (with the remaining 5 percent comprising nitrogen, argon, oxygen, and carbon monoxide). Back in the day, Mars had a thicker atmosphere. One theory suggests that the atmosphere became thinner due to a giant impact that stripped the atmosphere away. Still, some atmosphere is better than no atmosphere. The atmosphere on Mars is thick enough to support weather but not thick enough to prevent the mercury from dipping down to −60 °C at the equator and −125 °C near the poles (it can be as comfortable as 20 °C at the equator at midday).

■ **Table 1.3** Select properties of Mars

Planetary data for Mars	
Mean distance from Sun	227,943,824 km (1.5 AU)
Martian year (sidereal period of revolution)	686.98 Earth days
Mean orbital velocity	24.1 km/sec
Equatorial radius	3,396.2 km
Surface area	1.44×10^8 km^2
Mean surface gravity	371 cm/sec^2
Escape velocity	5.03 km/sec
Rotation period (Martian sidereal day)	24 hr 37 min 22.663 sec
Martian mean solar day (sol)	24 hr 39 min 36 sec
Mean surface temperature	−63 °C
Typical surface pressure	0.006 bar

1

Radiation is still a problem, as is dust, which is comprised of oxidized iron dust. But first, the radiation. The Martian surface receives 30 μSv per hour during solar minimum and about twice that during solar maximum [13, 14]. If astronauts were to spend 3 hours every 3 days outside their habitat, their exposure would be about 11 mSv per year. If they shield their habitat with 5 meters of soil, they will be afforded about the same protection as on Earth. We'll return to the problem of designing extraterrestrial life support later, but before we do, we'll learn a bit more about human physiology.

Key Terms
- Deoxyribonucleic acid (DNA)
- Electromagnetic (EM)
- Galactic Cosmic Radiation (GCR)
- International Space Station (ISS)
- Low Earth Orbit (LEO)
- Linear Energy Transfer (LET)
- Micrometeoroids and Orbital Debris (MMOD)
- Relative Biological Effectiveness (RBE)
- Ribonucleic acid (RNA)
- Solar Particle Event (SPE)
- SpaceShipOne (SS1)
- SpaceShipTwo (SS2)
- Space Surveillance Network (SSN)

❓ Review Questions
1. What is the difference between the lithosphere and the hydrosphere?
2. What is meant by the term *biomes*?
3. List three primary characteristics of the carbon cycle.
4. What is GCR?
5. What is the difference between LET and RBE?
6. Explain the significance of the Armstrong Line in the context of human physiology.
7. List three primary characteristics of the phosphorus cycle.
8. List the six layers of the atmosphere.

References

1. Cucinotta, F. A., Chappell, L., & Kim, M. Y. (2013). *Space radiation cancer risk projections and uncertainties–2012* (NASA Technical Paper 2013-217375, NASA STI Program, Hampton).
2. National Aeronautics and Space Administration. Space Radiation Analysis Group, Johnson Space Center. Washington, DC: The Administration; 2008.
3. National Council on Radiation Protection and Measurements. (2000). *Radiation protection guidance for activities in low-earth orbit*. Bethesda: The Council. Report no. 132. ISBN 0-929600-65-7.

4. DeLombard, R., Hrovat, K., Kelly, E., et al. (2004). *Microgravity environment on the International Space Station.* Washington, DC: National Aeronautics and Space Administration. Report no. NASA/TM—2004-213039.

5. Rodriguez, H. M., & Liou, J. C. (2008). Orbital debris: Past, present, and future. In: *Proceedings of American Institute of Aeronautics and Astronautics (AIAA) Annual Technical Symposium, 2008 May 9.* Houston, Webster: American Institute of Aeronautics and Astronautics.

6. Soares, C., Mikatarian, R., Schmidl, R., et al. Natural and induced space environments effects on the International Space Station. In: *Proceedings of the 56th International Astronautical Congress, 2005 October 17–21.* Fukuoka, Japan. IAC-05-B4.2.07.

7. Eckart, P. (1999). *Space flight life support and Biospherics.* Dordrecht: Kluwer Academic Publishers/Torrance: Microcosm Inc.

8. Nicogossian, E. A., Huntoon, C. L., & Pool, S. L. (Eds.). (1994). *Space physiology and medicine* (pp. 167–193). Philadelphia: Lea and Febiger.

9. National Council on Radiation Protection and Measurements. (1989). *Guidance on radiation received in space activities.* Bethesda: The Council. Report no. 98. ISBN 0-929600-04-5.

10. NCRP Report No. 98: Guidance on radiation received in space activities. (1989, July 31). Bethesda: National Council on Radiation Protection and Measurements.

11. Silberberg, R., Tsao, C. H., Adams, J. H., Jr., & Letaw, J. R. (1985). Radiation transport of cosmic ray nuclei in lunar material and radiation doses. In W. W. Mendell (Ed.), *Lunar bases and space activities of the 21st century* (pp. 663–669). Houston: Lunar & Planetary Inst.

12. Land, P. (1985). Lunar base design. In W. W. Mendell (Ed.), *Lunar bases and space activities of the 21st century* (pp. 363–373). Houston: Lunar & Planetary Inst.

13. Newman, D. J. (2000). Life in extreme environments: How will humans perform on Mars? *Gravitational and Space Biology Bulletin, 13,* 35–47.

14. Zeitlin, C., Hassler, D. M., Cucinotta, F. A., Ehresmann, B., Wimmer-Schweingruber, R. F., Brinza, D. E., Kang, S., Weigle, G., Böttcher, S., Böhm, E., Burmeister, S., Guo, J., Köhler, J., Martin, C., Posner, A., Rafkin, S., & Reitz, G. (31, May 2013). Measurements of energetic particle radiation in transit to Mars on the Mars. *Science Laboratory Science, 340*(6136), 1080–1084.

Suggested Reading

Finckenor, M., & de Groh, K. (2015). *Space environmental effects.* A mini-book published by NASA. Just 40 pages and available online at: https://www.nasa.gov/sites/default/files/files/NP-2015-03-015-JSC_Space_Environment-ISS-Mini-Book-2015-508.pdf

Thirsk, R., Kuipers, A., Mukai, C., & Williams, D. (2009, June 9). The space-flight environment: The International Space Station and beyond. *CMAJ: Canadian Medical Association Journal, 180*(12), 1216–1220. https://doi.org/10.1503/cmaj.081125. Epub 2009 Jun 1. A good reference written by astronauts.

Space Physiology and Psychology

Credit: Jim Wilkie

© Springer Nature Switzerland AG 2020
E. Seedhouse, *Life Support Systems for Humans in Space*,
https://doi.org/10.1007/978-3-030-52859-1_2

Contents

2

🏛 Learning Objectives

After reading this chapter, you should be able to:

- Describe the mechanism of bone loss in astronauts
- Describe the role of the osteoclasts and osteoblasts
- Explain what is meant by the term *bone remodeling*
- State the rate of loss of bone mass density in astronauts during long-duration missions
- List and describe the efficacy of two countermeasures to bone loss
- List the three types of muscle
- Explain the mechanism of muscle atrophy in astronauts, and explain to what degree muscle atrophy correlates with loss of muscle contraction
- Describe the sliding filament theory in the context of muscle atrophy and exercise
- Explain why some astronauts suffer from space motion sickness
- Describe the function of the semicircular canals and the otolith organ
- Describe the theory of vision impairment intracranial pressure (VIIP)
- Explain what is meant by intracranial pressure (ICP) in the context of VIIP
- Explain what is meant by posterior globe flattening
- Describe the mechanism of fluid shift when astronauts arrive on orbit
- Explain what is meant by the term *diuresis*
- Describe three key characteristics of the space food system
- Explain what is meant by the terms *overview effect*, *asthenia*, and *salutogenesis*
- List five "select-in" psychosocial characteristics for long-duration missions
- Describe the difference between galactic cosmic radiation (GCR) and solar particle events (SPE)
- Describe the effects of long-duration exposure to radiation on astronaut health
- Describe the effects of radiation on the central nervous system
- Explain what is meant by the term *altered neurogenesis*
- Describe the mechanism of radiation-induced oxidative damage
- Explain what is meant by the term *osteoradionecrosis*

Introduction

No textbook about spaceflight life support systems can be written without a basic overview of physiology. After all, the life support system is what keeps astronauts alive. To fully understand the intricacies of these engineering systems, it is necessary to first understand the physiological systems. To that end, ▶ Chap. 2 delves into basic physiology, with some psychology thrown in. One aspect of physiology emphasized in this chapter is that we humans can only operate within very, *very* narrow environmental parameters. For example, a partial pressure that is too high or too low can cause all sorts of problems. This human limitation represents a big challenge for life support system engineers. Another aspect highlighted is the adaptive timeline across physiological systems. Some systems, such as the fluid system, adapt to microgravity after 4–6 weeks, but other systems, such as the skeletal system, have no clinical horizon, so this is a good place to start.

Bone Loss

Bone plays several important roles. First, it serves as a structure that supports the body. Second, it stores calcium. Third, it produces blood. Bone is also very dynamic. This is because of a process known as bone remodeling [1, 2]. This process relies on the activity of two very important bone cells: *osteoblasts*, which build up bone, and *osteoclasts*, which break down bone (bone resorption). On Earth, your skeleton undergoes a tremendous amount of loading. Those of you who wear Fitbits will know that you take between 8000 and 10000 steps every day just performing daily activities. That loading is detected by your brain and signals those osteoblasts and osteoclasts to go to work (◘ Fig. 2.1).

Bone Loss in Spaceflight

The result of this process is a healthy skeleton that is fracture-resistant. But in space, almost all that loading is removed. Your brain detects that reduction in load and adapts accordingly. Less load means that strong bones are no longer needed, so the process of bone remodeling changes. The action of the osteoblasts is decreased and the action of the osteoclasts is increased. The result is a reduction in bone mass density of between 1.0 and 1.2 percent – every month. That is ten times the rate at which osteoporosis patients lose bone.

This is catastrophic (◘ Fig. 2.2). Losing that amount of bone mass density increases fracture risk two to three times. Further, the body not only alters the responses of the osteoblasts and osteoclasts, but it also alters the body's calcium balance. Calcium is a key bone-building material. But in microgravity, it is not necessary to have strong bones, so the body decides to get rid of calcium to the tune of ~250mg per day [2, 3]. Not only does calcium excretion result in weaker bones,

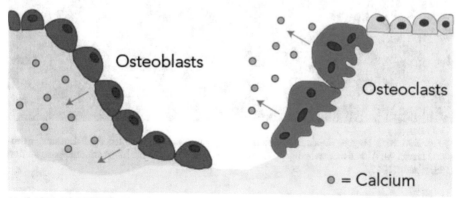

◘ **Fig. 2.1** The action of the osteoblasts and osteoclasts. Credit: NASA

2

Osteoporosis

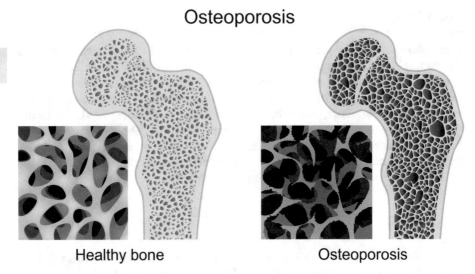

Healthy bone   Osteoporosis

Fig. 2.2   Osteoporosis. Credit: NASA

Fig. 2.3   Nick Hague, Russian cosmonaut Alexey Ovchinin, and United Arab Emirates astronaut Hazzaa Ali Almansoori relaxing after their increment on board the International Space Station (ISS). Credit: NASA

but it also increases the risk of kidney stones, because that calcium has to be routed through the kidneys before being excreted [4, 5, 6, 7]. The worst is still to come.

Take a look at ◘ Fig. 2.3. That is a photo of NASA astronaut Nick Hague, Russian cosmonaut Alexey Ovchinin, and United Arab Emirates astronaut Hazzaa

Ali Almansoori. They have just returned from an ISS increment lasting several months. Notice anything particular in this image (apart from the fact they're on their cellphones)? They are horizontal. Why? Because their bones (and muscles) are very weak after having spent so long in space. And those people milling around the crew? Some of them are flight surgeons entrusted with the care of the crew. Shortly after landing, the crew are whisked away to their respective agencies and embark on a dedicated and personalized rehabilitation program to regain lost muscle and bone. After about 90 days, most astronauts have recovered their muscle. But bone is another story. There have been some astronauts who, 10 years following their 6-month increment on the ISS, did not completely recover their bone loss. Some astronauts recover their bone mass relatively quickly (3–4 years), while other astronauts take longer.

Another aspect of this bone loss is bone mass density and architecture. Take another look at ◘ Fig. 2.2. You see those rod-shaped structures? Those help the architecture of the bone. The better the arrangement of those rods and plates, the better your architecture and – theoretically – the stronger your bones. But another element that has a bearing on bone strength is bone density. You can have fantastic bone architecture, but if your bone density is low, then your bones will be weak and vice versa. The ideal scenario is to have both good bone density and good bone architecture.

In space, the body remodels bone in response to the load demands on the bone. Since those load demands are low, bone density is reduced, and those rods and plates are arranged to deal with the loads in microgravity. This leads to weaker bone architecture. In low Earth orbit (LEO), the effect is manageable, thanks to the limited time (only 6 months give or take) in orbit and thanks to the personalized rehab programs waiting for astronauts when they return to Earth. But a Mars mission will require astronauts to spend the best part of 12 months in deep space, plus a surface stay in a reduced gravity (0.38 of Earth's gravity) environment. Imagine a crewmember suffering a femoral fracture. How would the crew cope? *Could* they cope?

Take a look at ◘ Fig. 2.4. The image depicted in ◘ Fig. 2.4 is one way of dealing with a femoral fracture. The technique is known as *external fixation*. Imagine all the challenges of dealing with this in a reduced gravity environment! Bleeding, infection, reinfection, sepsis, 24/7 care, etc. So what can we do?

Countermeasures to Bone Loss

Countermeasures to bone loss include treadmill running and using the Advanced Resistive Exercise Device (ARED), which is depicted in ◘ Figs. 2.5 (a) and 2.5 (b).

As you can see in ◘ Figs. 2.5 (a) and (b), the ARED is a versatile exercise device, and as we shall see in ▶ Chap. 7, the device does a good job at mitigating the effects of bone loss [8, 9]. But even astronauts who exercise religiously still suffer 1.0–1.2 percent bone loss per month. Some solutions are also discussed in ▶ Chap. 7.

2

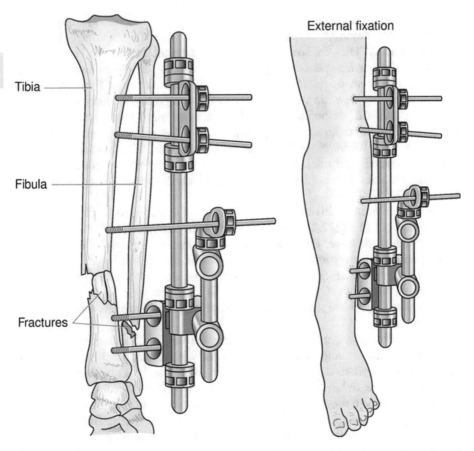

Tibia

Fibula

Fractures

External fixation

◘ Fig. 2.4   External fixation. Imagine a crew having to deal with this situation on the surface of Mars. Credit: The Free Dictionary

Muscle Loss

Muscle Physiology

There are three types of muscle in the body. *Smooth muscle*, also categorized as involuntary muscle, is found in organs and organ structures. *Cardiac muscle*, which is also involuntary muscle, is found only in the heart. *Skeletal muscle*, which is also categorized as voluntary muscle, is used for movement.

You have about 640 skeletal muscles in your body. These are anchored to bone by tendons. Also known as striated muscle due to its appearance under the microscope, skeletal muscle (◘ Fig. 2.6) helps support the body, assists in bone movement, and protects internal organs. It is divided into various subtypes. Type I, also known as *slow twitch* (red), is dense with capillaries, which means these muscles can carry a lot of oxygen and sustain lengthy periods of aerobic activity. Type II, also known as *fast twitch*, can sustain short bursts of activity.

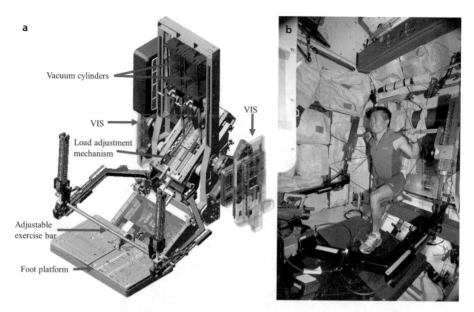

● **Fig. 2.5** **a** Schematic of the ARED system. Credit: NASA. **b** Japan Aerospace Exploration Agency astronaut Koichi Wakata, Expedition 38 flight engineer, works out on the Advanced Resistive Exercise Device (ARED) in the Tranquility node of the ISS. Credit: NASA

Structure of a Skeletal Muscle

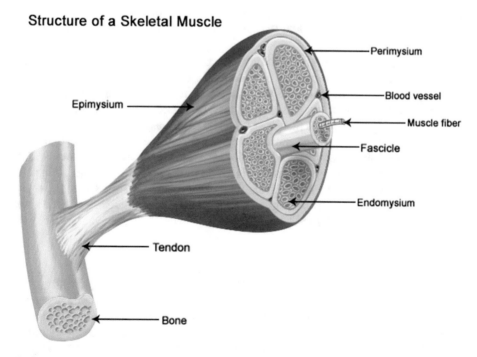

● **Fig. 2.6** Skeletal muscle structure. Credit: National Cancer Institute

2

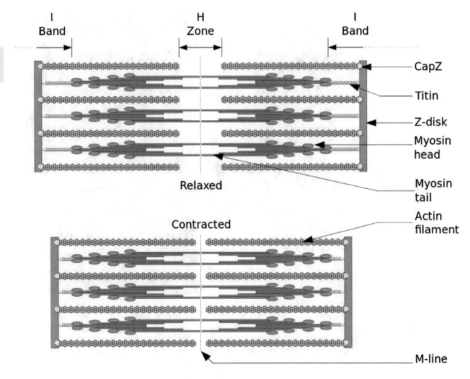

Fig. 2.7   The sliding filament theory. Credit: Richfield, David (2014). "Medical gallery of David Richfield." *WikiJournal of Medicine* 1 (2). DOI:10.15347/wjm/2014.009. ISSN 2002-4436. /CC BY-SA

Now some basic exercise physiology. Despite what you may hear in the gym, fat cannot be turned into fat, and fat cannot be turned into muscle. Impossible! Second, the number of muscle fibers cannot be increased no matter how hard you exercise. So, how do muscles get bigger? The muscle cells get bigger in a process known as *hypertrophy*. This is the opposite of *atrophy*, which is the term that describes the wasting away of muscles. In skeletal muscle, movement is achieved by contraction stimulated by nerve impulses at a site known as the neuromuscular junction. Energy for muscle movement is found in the form of glycogen, which is stored in the muscles and in the liver. When exercising, the muscles contract by actin and myosin filaments sliding over one another, a mechanism known as the *sliding filament theory* (Fig. 2.7).

Muscle Atrophy in Space

On Earth, muscle tone is maintained by means of exercise. Even sedentary people will take 8000 to 10,000 steps every day, which is usually sufficient exercise to maintain some muscle tone. But in space, muscles don't get used nearly as much as on Earth. The result is that muscles begin to atrophy at a rapid rate. In just 6 months, astronauts will lose about 25 percent of their total muscle mass [8, 9].

But this loss is not evenly distributed. A larger proportion of muscle loss is in the load-bearing muscles (your big leg muscles) and the balance muscles (e.g., those muscle groups that support your spine). As astronauts spend more and more time in space, their muscle cells shrink and become smaller. And as those muscles become smaller, they become weaker, which means astronauts cannot exert as much force.

Think operationally. Think of all those images and videos we've been bombarded with of astronauts going about their business bouncing around on the surface of Mars – astronauts building outposts and exploring the surface during lengthy EVAs. Do you really think they will be able to do these tasks in such a deconditioned state? Don't forget that these astronauts have lost a quarter of their muscle mass, which includes their cardiac muscle mass. This means the heart is much less efficient, leading to significantly reduced exercise capacity [10]. And as mission time increases, this will only get worse. So again, what can be done? Once again, countermeasures. Treadmill running and using the ARED help [8, 9]; we'll discuss these more in ▶ Chap. 7.

Neurovestibular Effects

Your vestibular system (◘ Fig. 2.8) provides information about motion, head position, and spatial orientation, and it is located in the inner ear. It comprises several structures that help you balance and maintain equilibrium and posture. The *semicircular canals* are arranged at right angles to one another and provide information about angular acceleration (pitch, roll, and yaw). The *otolith organs* provide information about linear acceleration. Another key element in the neurovestibular system is the hair cell (stereocilia). These hair cells are embedded in a gelatinous structure (called the cupula). When you move your head, these hair cells move/ shift, and information is carried from the hair cells to the brain (brainstem and cerebellum), which interprets the movement accordingly.

Space Motion Sickness

Disruption to the neurovestibular system, which may occur when exposed to microgravity, can cause symptoms such as nausea, vertigo, loss of balance, and vomiting [11]. The microgravity environment may not only be disruptive to the neurovestibular system but may also affect the visual system. Why? For years, we accumulate sensory information via our neurovestibular system and our visual system, and all this information is stored in a repository for sensory information. When something conflicts with that repository of information, motion sickness may be the result. In fact, 60 percent of first-time astronauts suffer from space motion sickness (SMS).

Take look at ◘ Fig. 2.9. It shows NASA astronauts Ron Garan and Cady Coleman, ESA astronaut Paolo Nespoli, and Russian cosmonaut Alexander Samokutyaev of Expedition 27 posing in the Harmony node of the ISS. Imagine seeing this when you arrive on orbit! Fortunately, as you can see in ◘ Fig. 2.10,

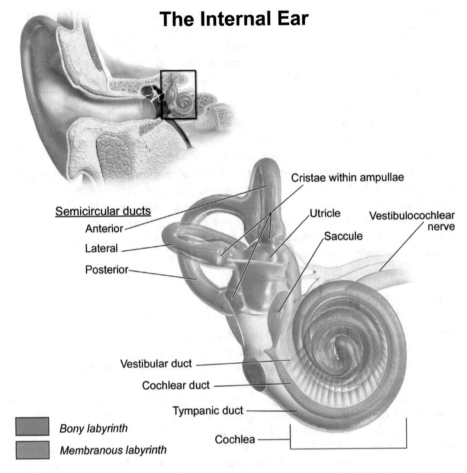

Fig. 2.8 Structure of the inner ear. Credit: Blausen.com staff (2014). "Medical gallery of Blausen Medical 2014." WikiJournal of Medicine 1 (2). DOI:10.15347/wjm/2014.010. ISSN 2002-4436. /CC BY-SA

the neurovestibular system adapts very quickly. It takes no longer 72 hours for this system to adapt.

Astronauts can take anti-motion sickness medication such as promethazine, although this doesn't always help. Autogenic feedback training is also helpful, but ultimately predicting who will be sick and who won't has been a mystery for as long as there have been astronauts.

Vision Impairment

By now we know that astronauts' bodies suffer in microgravity. Without effective countermeasures, muscles atrophy, bones shed calcium, and astronauts get sick. Eyesight may also be affected. We've known about vision impairment in astronauts

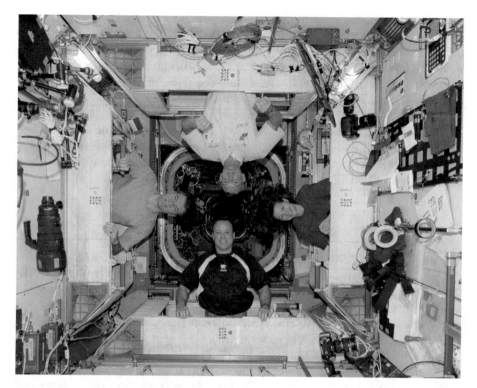

☐ Fig. 2.9 Ron Garan, Cady Coleman, ESA astronaut Paolo Nespoli, and Russian cosmonaut Alexander Samokutyaev of Expedition 27 posing in the Harmony node of the ISS. Credit: NASA

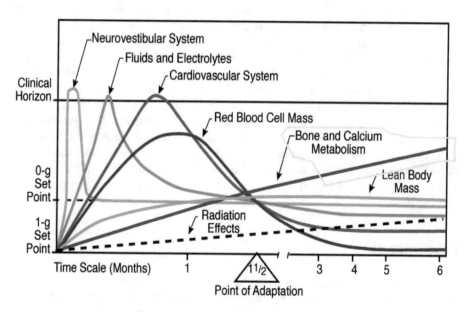

☐ Fig. 2.10 Timeline of adaptation for physiological systems. Credit: NASA

2

for some time, but the problem has only been put under the spotlight recently after some astronauts experienced severe eyesight deficiencies [12, 13]. Thanks to anecdotal reports by astronauts and a comparison of preflight and postflight ocular measures, microgravity-induced visual acuity impairments have now been recognized as a significant risk (you can read more about this in Springer's *Microgravity and Vision Impairments in Astronauts* written by yours truly). And this problem doesn't affect a minority of crewmembers: retrospective analysis of medical records has revealed that 29 percent of 300 Shuttle astronauts and 60 percent of space station astronauts have suffered some form of visual degradation [14]. That's a serious problem for an agency planning on sending astronauts back to the Moon and eventually Mars.

Theories

The problem has its own acronym – this is NASA, after all – and is referred to as the *visual impairment intracranial pressure (VIIP) syndrome*. Even though VIIP has only recently been identified, there has been significant research performed, so scientists are beginning to better understand the syndrome. The data shows that astronauts who suffer VIIP-related symptoms experience varying degrees of visual performance decrements. Some suffer cotton wool spot formation, while others may present with edema of the optic disc (◘ Fig. 2.11) [15, 16]. Other astronauts may suffer flattening of the posterior globe, while some may present with distension of the optic nerve sheath [15]. In short, there is a profusion of signs and symptoms, but the reason for the vision impairment still has researchers a little flummoxed.

One theory suggests that the changes in ocular structure and impairment to the optic nerve are caused by the cephalothoracic fluid shift that astronauts experience while on board the ISS [17, 18, 19]. It is theorized that some astronauts are more sensitive to fluid shift due to genetic and anatomical factors. In the course of conducting studies on the VIIP syndrome, researchers have focused on three systems: ocular, cardiovascular, and central nervous. These studies have revealed a variety of symptoms other than visual decrements, including increased intracranial pressure (ICP) and changes in cerebrospinal fluid (CSF) pressure [20]. But because preflight, inflight, and postflight data are relatively thin on the ground, it is very difficult to define why and how these symptoms occur. Inevitably, since the impact of VIIP is an operational concern, space agencies have increased preflight, inflight, and postflight monitoring of the syndrome to better characterize the syndrome and the risks.

Case Studies

To date, more than a dozen astronauts have suffered VIIP symptoms [13]. Some of these symptoms have persisted postflight and some haven't. Some astronauts experience quite severe symptoms, and for others, it is just a mild inconvenience. Very

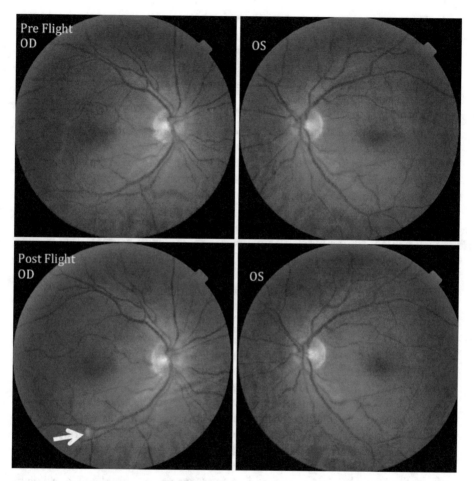

☐ **Fig. 2.11** Cotton wool spots. Credit: NASA

little is known about the etiology of symptoms, but researchers believe the afore-mentioned microgravity-induced cephalothoracic fluid shift and associated physi-ological changes may be implicated. To better investigate the syndrome, NASA's Human Health and Performance Directorate brought together a VIIP project team in 2011 to investigate the problem using data compilation, analysis, and multidis-cipline, cross-cutting collaboration. To give you an idea of the complexity of the issue facing the VIIP project team, the following is a brief account of three cases of the perplexing syndrome:

Case #1. This case occurred 3 months into an ISS mission when a crewmember informed the ground he was only able to see the Earth clearly if he used his reading glasses. For the remainder of his mission, there was no improvement in his condition but neither did the symptoms get worse. After his return to Earth, the astronaut noticed a gradual improvement in his vision, but his vision didn't return completely. The astronaut was subjected to fluorescein angiography, which revealed choroidal

2

folds. This astronaut was also subjected to magnetic resonance angiography (MRA) and magnetic resonance venogram (MRV) tests, but these were normal [13]. A more sensitive test using optical coherence tomography (OCT) was also performed. OCT is a noninvasive test that uses light waves to take images of the retina. The results showed increased thickening of the astronaut's retinal nerve fiber layer (NFL). It was one more variable in the VIIP puzzle.

Case #2. This astronaut noticed his vision altering after 2 months on orbit. His symptoms included a reduction in the near vision of his right eye and scotoma in his right temporal field. The scotoma affected the crewmember to the extent that he was unable to read 12-point font, which is close to what you're reading now. What was strange was that everything about this astronaut was normal: he didn't complain of any symptoms – headache, diplopia – that might normally be associated with such vision impairment [13]. Also, the environment on board the ISS was well within nominal margins during his stay – no excess carbon dioxide concentration (carbon dioxide levels are much higher on the ISS, and some scientists think this may cause VIIP) or toxic fumes. Furthermore, his preflight and postflight (correctable) visual acuity were the same, although fundoscopic examination revealed choroidal folds in the astronaut's right eye.

Case #3. This ISS astronaut's vision was affected so badly that his glasses had to be adjusted to enable him to read procedures [13]. By this time, NASA decided it is prudent to perform inflight ocular ultrasound exams to observe changes in the ocular health of its crew. To that end, this astronaut was subjected to ultrasound examinations, which revealed flattening of the posterior globe and dilated optic nerve sheaths. This case occurred way back in the day when the Shuttle was flying, so mission managers decided to fly a video ophthalmoscope, which also allowed remotely guided fundoscopic exams to be performed. Once the examinations had been performed, the images were sent to neuro-ophthalmological consultants, who decided that no treatment was required but suggested that fundoscopic and visual acuity exams be conducted once a month for the remainder of the mission. The examinations served a dual purpose: they provided neuro-ophthalmologists with a baseline to compare against other vision issues, and they served as a means of monitoring the ocular health of the crewmembers. Case #3 continued to be examined following his flight. 30 days postflight, another MRI was performed and revealed significantly dilated optic nerve sheaths, flattening of the posterior globe and thickened optic nerves [13]. At this stage, the crewmember was still affected by the same vision impairment he had experienced on orbit. Another examination, a lumbar puncture, performed 57 days postflight revealed normal CSF pressure. The astronaut, who became aware of visual changes 3 weeks into his mission, reported that his vision had remained static for the remainder of his stay on board the ISS [13]. He hadn't complained of headaches, visual obscurations, or diplopia, and yet several weeks postflight, he was still suffering vision impairment. More tests were conducted, including cycloplegic refraction and fundus photos, but these were all normal: no sign of choroidal folds or disc edema. Nothing.

Many cases of VIIP have shared the common signs of disc edema, posterior globe flattening, choroidal folds, and hyperopic shift. These signs closely align with those reported by terrestrial patients suffering from increased intracranial hypertension (Appendix II). These cases also shared other similarities, including optic nerve sheath distension and posterior globe flattening. These observations were revealed by MRIs performed post-mission. Despite myriad tests (summarized in Appendix II) and scans, a definitive etiology for the findings in the affected astronauts remained elusive. It was suggested that venous congestion in the eye, caused by cephalothoracic fluid shift, could have been responsible for elevated choroidal volume changes [18]. Given the commonality of that sign, it was also suggested that this could be a unifying mechanism.

Intracranial Pressure

The reason researchers were interested in examining those who suffered from intracranial hypertension was because the ophthalmic changes these terrestrial patients reported were so similar to those reported by astronauts [21, 22]. Researchers reasoned that the increased intracranial pressure (ICP) suffered by terrestrial patients might somehow explain the visual changes observed in astronauts. It was a good hypothesis, but slightly flawed. While ICP is reported by those suffering from intracranial hypertension and by astronauts suffering from VIIP, the mechanism of ICP is not the same in both groups. Those complaining of ICP here on Earth typically report much more severe symptoms than those on board the ISS. In addition to the pain of headaches, terrestrial ICP sufferers report cognition impairments and nausea and vomiting. At this point, it is worth emphasizing the numbers of patients we're dealing with: the terrestrial ICP problem has been observed in thousands and thousands of patients, whereas the VIIP syndrome has been reported in only a dozen or so astronauts. So, given the few "patients" on orbit, it's difficult to make correlations between terrestrial ICP and its space-based equivalent.

Cerebrospinal Fluid

Another complicating factor is that of the very, *very* small number of astronauts who have been studied; only a handful who reported ICP have had ICP measurements taken preflight and postflight. ICP measurements are usually taken by performing a lumbar puncture (sometimes referred to as a spinal tap). This procedure involves inserting a needle between two lumbar vertebrae to remove cerebrospinal fluid (CSF). When this procedure was performed postflight on astronauts who had reported VIIP symptoms, their CSF pressure was only mild to moderately above normal pressure. While this didn't significantly strengthen the hypothesis of increased ICP in flight, it did go some way to helping researchers understand the mechanisms implicated in the syndrome.

2

Optic Nerve Sheath Diameter

Another observation that bolstered the ICP hypothesis was measurements performed on astronaut's optic nerve sheath diameter (ONSD) and optical diameters (OD). These ultrasound postflight measurements revealed a distension of the ONSD, and since this is an indication of intracranial hypertension, the hypothesis of ICP being a symptom of VIIP gained some more traction [19, 21]. But only some. The findings were inconclusive [20], since postflight images were taken a long time after landing, in some cases months or years. Another drawback to what had seemed promising evidence of the ICP link was the fact there was no control group to compare the ONSD measurements with. At the time, there was a lot of talk about the VIIP problem being similar to intracranial hypertension, which caused some confusion as to the exact nature of the syndrome. In particular, the term *papilledema* was sometimes used to refer to the swelling of the optic disc observed in those reporting VIIP symptoms. But papilledema is simply the swelling of the optic disc caused by an increase in ICP, whereas in space the optic disc swelling could have been caused by ONSD or other effects (damage perhaps) on the optic nerve head.

The ONSD link supported the rationale for studying intracranial hypertension and its usefulness in understanding VIIP. However, it should be noted that while intracranial hypertension is well-characterized, the causes are poorly understood. The same is true for VIIP. For example, visual changes are not reported by all crewmembers who report VIIP symptoms, and in some crewmembers, these symptoms can be quite pronounced, whereas in others the symptoms may barely register. The point is that there seems to be little rhyme or reason or pattern to the symptoms. Part of the reason for this is the fact that so few crewmembers have been evaluated. To give you an idea of just how confusing the symptom pattern is, consider the case of one crewmember who reported no symptoms during or after his flight but whose postflight examination revealed the worst case of optic disc edema ever observed! This crewmember not only had Grade 3 edema in his right optic disc and Grade 1 in his left but also had a small hemorrhage inferior to the optic disc of his right eye, nerve fiber thickening, *and* increased CSF pressure! Having said that, this crewmember had no signs of choroidal folds or globe flattening. It was a head-scratching case that left many researchers puzzled.

Posterior Globe Flattening

Back to the ICP issue. Another confounding variable in the link between intracranial hypertension, ICP and VIIP, is the symptom of posterior globe flattening. On Earth, posterior globe flattening can be induced by ICP, but after treatment for intracranial hypertension, no one knows if globe flattening persists because there just haven't been any studies to follow up on the issue. This means that no one knows if pressure behind the optic disc results in permanent posterior globe flattening. It's just another hole of knowledge in the VIIP syndrome.

Microstructural Anatomical Differences

Perhaps ICP, and therefore VIIP, is a function of anatomical differences? After all, there are several structures in the eye that are affected by pressure, and differences in the morphology of these could result in some crewmembers suffering more symptoms than others. Let us take the *lamina cribrosa sclera* as an example. This structure is located between the optic nerve and the intraocular space, and it serves as a pressure barrier between that space and the retrobulbar CSF space that surrounds the retrobulbar part of the optic nerve [20].

On Earth, under normal conditions, the lamina cribrosa bulges slightly outward in the direction of the optic nerve. This is because intraocular pressure (IOP), which is 15 mmHg, is higher than intracranial pressure, which is 10 mmHg. But if these pressures are changed in any way, the lamina cribrosa will respond in a way that may affect vision. For example, if the structure is displaced excessively, the lamina cribrosa can squeeze adjacent nerve fibers, causing nerve damage and vision loss. This is essentially what happens in people with glaucoma. But this mechanism doesn't affect everyone equally. That's because a person's anatomy can make them more or less susceptible to syndromes and ailments. So if you happen to be an astronaut whose tissue elasticity is greater than most, chances are you may be less affected by pressure changes exerted on your lamina cribrosa. That's because the more elastic a tissue is, the lower the chances are that a significant pressure chance will have a damaging effect on the sensitive tissues in the eye and the brain. Equally, those who had an unfortunate genetic roll of the dice and who have less elastic tissues are probably at greater risk of ICP-induced visual changes. Remember that crewmember who reported no symptoms inflight but who was diagnosed with Grade 3 edema on his return? He probably experienced increased ICP, but because his vessels and tissues were more compliant than most, he didn't suffer any visual deficits.

Cardiovascular System

When considering the cardiovascular system and how it adapts to microgravity, this system, like so many physiological systems, remains poorly understood. The adaptive process is complex and involves many control mechanisms, such as the autonomic nervous system, cardiac functions, and peripheral vasculature.

Fluid Shift

The primary cause of and trigger for these adaptive processes is the headward fluid shift and redistribution of body fluids that occur in every astronaut on arrival on orbit (■ Fig. 2.12). The body has about 5 liters of blood in addition to other body fluids, such as interstitial fluid, found between the organs, and CSF, found in the spinal cord. When astronauts arrive on orbit, between 1.5 and 2.0 liters of this fluid moves from the lower extremities to the chest and head [22]. Not surprisingly,

2

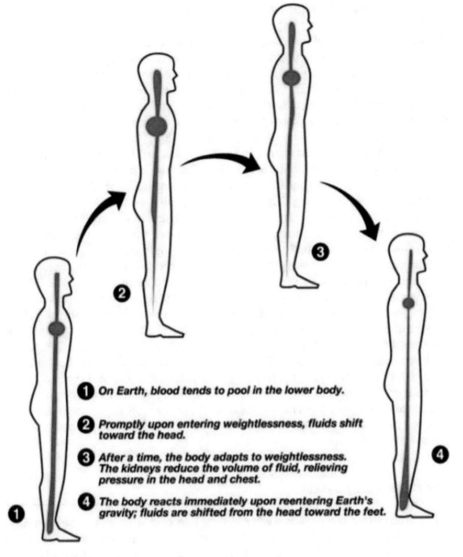

1 On Earth, blood tends to pool in the lower body.

2 Promptly upon entering weightlessness, fluids shift toward the head.

3 After a time, the body adapts to weightlessness. The kidneys reduce the volume of fluid, relieving pressure in the head and chest.

4 The body reacts immediately upon reentering Earth's gravity; fluids are shifted from the head toward the feet.

◘ Fig. 2.12   Fluid shift in microgravity. Credit: NASA

this causes various signs and symptoms, including facial puffiness, "bird-leg" syndrome, pounding headaches, and those vision problems mentioned earlier.

Diuresis

Remember the body will always try to adapt to the environment. This fluid shift triggers a series of adaptive processes. Inside your body, you have all sorts of sensors and receptors that send the brain information about temperature, electrolyte balance, and pressure. The pressure receptors are termed *baroreceptors*, and as long as fluid is maintained within certain thresholds, no action has to be taken. But when up to 2 liters of fluid is translocated from the lower to the upper body, this causes a spike in pressures that exceed thresholds. The baroreceptors send this information to the brain, and the brain decides that something must be done.

That something is to reduce pressure by getting rid of the excess fluid. This is done by triggering suppression of the renin-angiotensin-aldosterone system, which releases atrial natriuretic peptide, which ultimately results in *diuresis*, a physiological term meaning you have to visit the washroom frequently to urinate [22, 23, 24, 25]. Unfortunately, all this urination has a side effect of reducing plasma volume. About 55 percent of your blood is plasma and about 90 percent of your plasma is water. So, if you're urinating frequently, you are losing body water and hence blood volume. In fact, in the first 24 hours on orbit, astronauts lose 17 percent of their plasma volume, which equates to an overall reduction of about 10 percent of total blood volume [26, 27, 28, 29]. The body does its best to adapt, which takes about 6 weeks, although this adaptation is to microgravity, not one-G. On return to Earth, guess what happens? All that fluid that was in the upper body rushes to the lower body, causing orthostatic intolerance (inability to stand upright). Twenty-five percent of astronauts returning from space cannot stand upright for 10 minutes within hours of landing because of orthostatic intolerance. This is one of the reasons why NASA prohibits astronauts from driving 3 weeks postflight.

Nutritional Issues

Maintaining adequate nutritional intake during long stints on the ISS is important because astronauts must not only meet the usual terrestrial nutrient requirements but must also try to stave off the negative effects of being exposed to microgravity for so long. Also, as history has taught us, the success or failure of exploration missions (■ Fig. 2.13) has generally been determined by the degree to which nutrition was considered. Scurvy, anyone?

Although Shackleton, pictured in Ocean Camp next to Frank Wild in ■ Fig. 2.13, largely avoided scurvy, this nutrient deficiency led to more crewmember deaths in the Heroic Age of Antarctic Exploration than all other causes of their death combined. Arguably, nutrition will be even more important to astronauts. At least Shackleton and his crew could kill penguins and seals when their rations ran out – for beyond-Earth orbit missions, there will be no such resupply. So, food provision will be critical and meticulously planned to ensure optimal nutrition. This planning is a challenge because there are so many physiological changes (space motion sickness, fluid shifts, etc.) and spacecraft environment (lack of ultraviolet light, high levels of carbon dioxide, etc.) variables that can affect nutritional requirements.

Space Food System

The ISS food system (■ Fig. 2.14) provides menus on an 8–16-day cycle. Food items are supplied by NASA, CSA, ESA, JAXA, and the Russian Space Agency. Food is packaged in single-serving containers. The food can be in natural form, or

2

□ **Fig. 2.13**   The legendary Shackleton with his men on the ice. Credit: Frank Hurley

□ **Fig. 2.14**   The space food system on the ISS. Yum! Credit: NASA

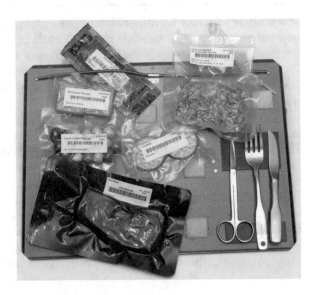

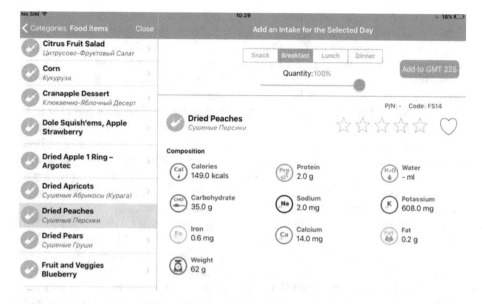

Fig. 2.15 Daily nutritional intake report. Credit: Carrier

it can be thermostabilized, dehydrated, and/or irradiated. Every day, crewmembers record their dietary intake using a Food Intake Tracker (FIT) app (■ Fig. 2.15) by simply scanning the barcode of the foods to be eaten. The food the crew eats is based on specific menus, but the astronauts can also bring bonus foods (Sunita Williams is a big fan of fluffernutter, which is a sandwich made with peanut butter and marshmallow crème, usually served on white bread). Analysis of the nutrients consumed by the crew is conducted by Johnson Space Center's Food Analytical Laboratory, which generates details about food intake, for example, how much protein, iron, and calcium each crewmember is eating.

Nutritional Countermeasures

All this recording of data may seem excessive, but missions have revealed that astronauts do not always eat as much as they should. Apollo astronauts only ate about 65 percent of what they should have, while Shuttle crews fared little better, eating only about 70 percent of what was recommended. Not surprisingly, this lack of food intake was associated with weight loss and lack of energy. And if you are less energetic, that has a negative impact on not only performing exercise countermeasures but also on bone and muscle loss.

2

Another reason for fastidiously monitoring nutrition intake is to track biochemical differences preflight, inflight, and postflight. Remember the vision problems that were discussed earlier? Data suggests that carbon dioxide levels, which are much higher on ISS than on Earth, may increase blood flow to the head and compound vision issues. And biochemical evidence has indicated a link between circulating metabolites of the one-carbon metabolism pathway (such as homocysteine and methylmalonic acid) and these vision issues. This data has also indicated that concentrations of serum folate and certain one-carbon intermediates are related to changes in refraction following increments on ISS. Sodium has also been implicated in the vision issue. Sodium levels are very high: the average on Earth is >5g sodium per day, whereas astronaut intakes are around 12 to 13 g per day in the space food system. All these nutritional links to just one physiological issue underline the importance of tracking nutrition during missions.

Here's another issue: bone loss. Bone loss represents a critical challenge to astronaut health and is probably a sure-fire mission killer for any Mars mission. Data has suggested that nutritional manipulation may have some measurable effect on increasing bone formation. For example, increased remodeling rate was observed in crewmembers who maintained vitamin D status. Another bone loss countermeasure may be omega-3 fatty acids, but more research must be conducted to support this. But while excesses of certain elements of the space food system can help astronauts, excess intake of other elements may harm crewmembers. For example, the highest tolerable intake of iron is 45 mg per day, but some crewmembers have reported mission averages in excess of this. Excess iron content in food can cause oxidative damage.

Psychosocial Support

Astronauts possess a wide-ranging repertoire of behavioral competencies that help them function effectively in a multicultural environment. This repertoire is critical because a spacecraft is an environment in which faults cannot be tolerated. To ensure all astronauts have these skillsets, space agencies apply very specific "select-out" and "select-in" criteria during selection (see below). And once selected, astronauts complete extensive preflight training to develop "expeditionary behavior," which comprises space-related psychosocial skills designed to ensure mission success. In addition to all this preparation, to ensure missions proceed smoothly, there is a ground-based complement of staff who provide behavioral support via videoconferences.

» Single men, perfect health, considerable strength, perfect temperance, cheerfulness, ability to read and write English, prime seamen of course. Norwegians, Swedes and Danes preferred. Avoid English, Scotch and Irish. Refuse point-blank French, Italians and Spaniards. Pay to be Navy pay. Absolute and unhesitating obedience to every order, no matter what it may be.

» Captain De Long's crew requirements for the Jeanette Arctic Expedition 1879–1881

» The CSA is seeking outstanding scientists, engineers and/or medical doctors with a wide variety of backgrounds. Creativity, diversity, teamwork, and a probing mind are qualities required to join the CSA's Astronaut Corps. To withstand the physical demands of training and space flight, candidates must also demonstrate a high level of fitness and a clean bill of health.

» Canadian Space Agency's Astronaut Recruitment Campaign announcement, 2008

Crew Selection

Still, the process of selecting astronauts is a challenging task. Interpersonal dynamics and difficulties, crew performance breakdown, and human interaction and performance in a confined and isolated environment are just a few factors that must be considered. Additional selection criteria such as communication competence and intercultural training also have a decisive impact on future mission success. It is for these reasons that the psychological aspect is often identified as one of the more problematic life support issues.

Nothing could be further from the truth. There is a wealth of information about selecting crews for arduous expeditions. Take Shackleton. Back in 1913, nearly 5000 men applied for the 27 jobs available on the great explorer's Imperial Trans-Antarctic Expedition. By comparison, NASA received just 3654 applications for its 2009 astronaut selection campaign – and this was in the age of the Internet, remember. Shackleton personally interviewed each candidate he felt had potential. While he obviously had to have crewmembers with sailing and scientific skills, he also wanted people who had enthusiasm and optimism to help cope with expedition demands. Fortunately, Shackleton (◘ Fig. 2.16) had an eye for talent and knew how to build a team that could survive just about anything. Each and every crewmember was selected to do a specific job and do it well, which is probably why Shackleton's team survived for more than 2 years in Antarctica when all seemed lost.

In common with long-duration missions to ISS, Shackleton's expedition was a dangerous one, and the success of his mission depended on a good team. His ideal crewmember had to be qualified for work on board the *Endurance*, but they also had to have special qualifications to deal with the extreme (polar) conditions. Another vital quality was the ability to live together in harmony for a very, *very* long time without outside communication. Organizations that have studied how Shackleton survived find that even though his mission failed, every man survived the impossible odds because Shackleton had picked a good team and had made sure each member understood his role. Also, Shackleton knew how well the rigors of Antarctic exploration would test the spirit of his men, so he

2

◘ Fig. 2.16 Probably the greatest explorer who ever lived. Sir Ernest Shackleton. Credit: National Library of Norway

was careful to look for character and not just competence. Technical qualifications were an asset, but he placed a greater emphasis on a positive attitude and a light-hearted nature.

For example, when Shackleton interviewed Reginald James, who became the expedition's physicist, he asked whether James could sing! Alexander Macklin, a surgeon, won a place on the expedition when, in response to Shackleton's inquiry about why Macklin wore glasses, Macklin replied, "many a wise face would look foolish without spectacles." Then there was the selection of his leaders. With over two dozen men to command, Shackleton understood the value of having loyal and strong leaders, which is why he chose Frank Wild as his second-in-command. Wild was a veteran of Antarctic exploration who had more than proven his mettle and his compatibility with Shackleton on Shackleton's 1907 expedition. Likewise, Thomas Crean, Shackleton's second officer, had proven his strength and discipline in his service with Shackleton on a 1901 expedition.

- **Shackleton's ten guidelines for choosing crewmembers:**
  1. Start with a solid crew you know from previous expeditions or who come recommended by those you trust.
  2. Your second-in-command is the most important hire. Pick one who complements your management style, shows loyalty without being a yes man, and can work with others.
  3. Hire those who share your vision.
  4. Weed out those who are not prepared to do mundane or unpopular jobs.

5. Go deeper than job experience and expertise. Ask questions that reveal a candidate's personality, values, and perspective on work and life.
6. Surround yourself with cheerful, optimistic people. Not only will they reward you with the loyalty and camaraderie vital for success, but they will stick by you when times get tough.
7. Applicants hungriest for the job are apt to work hardest to keep it.
8. Hire those with the talents and expertise you lack.
9. Spell out clearly to new employees the duties and requirements of their jobs.
10. Help your crew do top-notch work.

Astronauts with the Wrong Stuff

While Shackleton's guidelines identify many characteristics that also apply to those being considered for ISS missions, those doing the selecting have to apply slightly more rigorous processes than Shackleton did to avoid selecting astronauts with the "wrong stuff".

(Panel 1) Crash and Burn: The Cautionary Tale of Lisa Nowak

It took NASA astronaut, Lisa Nowak (◘ Fig. 2.17) 12 days, 18 hours, 37 minutes and 54 seconds to secure her place in one of the world's most elite clubs when she flew aboard the Shuttle Discovery during STS-121 in July 2006. It took her about 14 hours to destroy it. That was how long it took the 43-year old mission specialist to drive the 1500 kilometers from Houston, Texas, to Orlando, Florida, carrying with her a carbon-dioxide powered pellet gun, a folding knife, pepper spray, a steel mallet and $600 in cash. Nowak had discovered that Colleen Shipman, a US air force captain, was flying in from Houston to Orlando that night and Nowak wanted to be there to 'scare her' into talking about her relationship with the man at the centre of a love triangle. That man was Bill Oefelein, who underwent astronaut training with Nowak.

Shipman allegedly saw Nowak, whom she had never met before, wearing a trench-coat, dark glasses and a wig, following her on a bus from an airport lounge to a car park. Afraid, she hurried to her car. She could hear running footsteps behind her and as she slammed the door Nowak slapped the window and tried to pull the door open. 'Can you help me, please? My boyfriend was supposed to pick me up and he is not here,' Nowak was alleged to have pleaded. When Shipman said she couldn't help, the astronaut started to cry. Shipman wound down her window at which point Nowak discharged the pepper spray. Shipman drove off, her eyes burning, and raised the alarm. Nowak was subsequently charged with attempted first-degree murder in what quickly became the most bizarre incident involving any of NASA's active-duty astronauts.

To say the group to which Nowak belonged (her assignment to the agency was terminated by NASA on March 8, 2007) is select is an understatement. Up

Fig. 2.17 Lisa Nowak. Credit: NASA

to 2007, NASA had selected just 321 astronauts since the US agency began preparing to go into space in 1959. Nowak had been subjected to NASA's rigorous screening process and trained for years to cope with the stress of space-flight. Like all astronauts, Nowak had been subject to extensive psychiatric and psychological screening, all of which made her behavior incomprehensible.

To many, the Nowak scandal called to mind every bad science fiction movie where they send unstable characters into space. Others argued that NASA should have noticed the signs of Nowak's unraveling. These people might have had a point, but you have to remember that people in highly stressful jobs are generally over-achievers who put a high value on per-

formance and a low value on self-care beyond that required to perform the job. These types – astronauts – do a great job ignoring and denying signs of fatigue, either physical or psychologi-cal - just like polar explorers. Instead, they assume a machine-like thought process to deal with any problems. But the human brain isn't just a thinking machine, it is also the seat of emotions, and the suppression of emotions plays out in the battlefield of the subcon-scious mind. That suppression and the associated physical and psychological damage eventually surfaces in skewed thought processes and actions, which is exactly what happened to Nowak.

Nowak's drama played out in an air-port parking lot. Imagine a comparable scene on a spaceship en-route to Mars!

As tragic as astronaut Lisa Nowak's breakdown was, NASA has every reason to be thankful that it didn't occur during a space mission and every reason to worry about how it will avoid such scenarios during missions to the Moon and Mars. Although research into genomic screening, brain scans, and biometric monitoring may help avoid or at least avert mental breakdowns, the Lisa Nowak incident (Panel 1) highlights the fact that the greatest threat to any long-duration mission might be the astronauts themselves!

Just like trying to predict whether a crewmember had the right stuff for a polar expedition, trying to guess whether an astronaut will be vulnerable to psychiatric or psychosocial problems during multi-month missions remains an inexact science at best. Current screening involves a standard battery of tests that are administered to collect psychological information, as well as successive 2-hour interviews [30]. The first interview is conducted by a psychiatrist and a psychologist and the second with a psychiatrist alone. Standard tests (which the Canadian Space Agency also employs to select its astronauts) include the Minnesota Multiphasic Personality Inventory and the Personality Characteristics Inventory, which are used to identify "right stuff," "no stuff," and "wrong stuff" characteristics [31, 32]. What constitutes the "right stuff" for a long-duration space odyssey includes a history of emotional stability and little sign of depression or neuroticism. These long-duration crewmembers tend to be socially adept introverts who get along well with others but don't need other people to be content. Another important characteristic is a high toleration for lack of achievement, which makes sense when you think about the sheer length of the mission. In common with Shackleton, Amundsen, and company, this group of astronauts need to be prepared for changes of plan, contingencies, and the possibility that goals won't be achieved.

Confinement

Another reason the doom and gloom merchants like to highlight the psychological aspect as being the weak link in manned spaceflight is the fact that astronauts work in isolated, confined, and extreme (ICE) environments. Working in these environments, aforesaid doom and gloom merchants argue, will cause astronauts to lose their minds, turn on each other, and perhaps even come to blows.

This is fantasy. Let us look at a real-life ICE experience to prove this point (two more are provided in Appendix III). In 1893, Fridtjof Nansen sailed to the Arctic in the *Fram* (◘ Fig. 2.18), a purpose-built, round-hulled ship designed to drift north through the sea ice. Nansen's theory was inspired by the voyage of the *Jeannette*, which foundered northeast of the New Siberian Islands and was found on the southwest coast of Greenland after having drifted across the Polar Sea. Nansen reckoned the Polar current's warm water was the reason for the movement of the ice. But after more than a year in the ice, it became apparent that *Fram* would not reach the North Pole. So Nansen (◘ Fig. 2.19), accompanied by Hjalmar Johansen, continued north on foot when the *Fram* reached 84° 4′ North. It was a bold move, as it meant leaving the *Fram* not to return and a return journey over drifting ice to the nearest known land 800 kilometers south of the point where they started.

2

◨ **Fig. 2.18**   The Fram, one of the most famous exploration vessels ever built. From Amundsen, Roald: *The South Pole, Vol. I*, first published by John Murray, London 1912

Nansen and Johansen started their journey on 14 March 1895 with three sledges, two kayaks, and 28 dogs. On 8 April 1895, they reached 86° 14′ N, the highest latitude ever reached at that time. The men then turned around and started back but they didn't find the land they expected. On 24 July 1895, after using their kayaks to cross open leads of water, they came across a series of islands where they built a hut (◨ Fig. 2.20) of moss, stones, and walrus hides. Here they spent 9 mostly dark months, spending up to 20 hours out of every 24 sleeping, waiting for the daylight of spring. They survived on walrus blubber and polar bear meat. In May 1896, Nansen and Johansen decided to strike out for Spitsbergen. After traveling for a month, not knowing where they were, they were delivered from their endeavors through a chance meeting with Frederick George Jackson, who was leading the British Jackson-Harmsworth Expedition, which was wintering on the island. Jackson informed them that they were on Franz Josef Land. Finally, Nansen and Johansen made it back to Vardø in the north of Norway.

◘ **Fig. 2.20** Artist Lars Lorde's depiction from a photograph by Nansen of the dugout hut where Nansen and fellow explorer Hjalmar Johansen spent the winter of 1895–1896. From Nansen's 1897 book *Farthest North*

Salutogenesis

2

Another negative that the doom and gloom merchants like to focus on is boredom. Personally, I've never heard of any astronaut complaining of being bored. In fact, in all the astronaut autobiographies written, the focus is mostly on the positive. Psychologists even have a term for how astronauts spin the positive aspects of being in such an isolated and extreme environment: *salutogenesis*. It's a term coined by Aaron Antonovsky, a professor of medical sociology, and it is intended to convey the idea that under certain conditions, stress is beneficial and health-promoting, not pathogenic or destructive to health. As you can imagine, polar explorers experienced all sorts of negative effects as they struggled to cope with isolation, deprivation, and extreme conditions. But on the flipside, the elation of having coped with so much successfully brought positive benefits. So explorers tended to enjoy the experience and had positive reactions to the challenges of the environment. Not only that, but this unique group of individuals thrived on the feeling of having successfully overcome these challenges. In their diaries, they routinely refer to the beauty and grandeur of the land, ice, and sea, the camaraderie and mutual support of the team, and the thrill of facing and overcoming the challenges of the environment, which is probably why so many signed up for repeat expeditions.

But space agencies and space psychologists are still fixated with the deleterious effects of long-duration missions and their countermeasures, and scant attention has been paid to the beneficial effects of such an endeavor (◘ Table 2.1), which is a shame, because polar exploration has shown that individuals who adapt positively to an inhospitable or extreme environment can derive benefits from their experiences, including an initial improvement in mental health as a crewmember adapts to the environment.

◘ **Table 2.1**   Salutogenic after-effects of polar expeditions (many of which apply to astronauts)

Sense of personal achievement

Striving toward important goals

Courage, resoluteness, indomitability

Excitement, curiosity

Increased self-esteem

Hardiness, resiliency, coping

Improved health

Group solidarity, cohesiveness, shared values

Increased individuality, reduced conformity

Ability to set and achieve higher goals and changes in thinking

Despite some researchers choosing to ignore the salutogenic effects of space-flight, these effects have been observed during most missions. Astronauts report positively about friendship and the cohesiveness among the crew, satisfaction in jobs well done, pride in having been chosen to fly in space, and an appreciation of the beauty of Earth from space. This has been coined the *overview effec*t (I strongly recommend the following video that describes this phenomenon: ► https://www.youtubc.com/watch?v=CHMIfOecrlo). In fact, the current trend in memoirs written by spacefarers is to refer to positive emotions three times as often as to negative ones, a good recent example being Chris Hadfield's *An Astronaut's Guide to Life on Earth: What Going to Space Taught Me About Ingenuity, Determination, and Being Prepared for Anything.* Astronauts' autobiographical accounts routinely mention trust in others, autonomy, initiative, industry, strong personal identity, and a conviction that their life makes sense and is worthwhile. These astronauts were confident about their emotional stability and coping abilities and viewed themselves as active agents in dealing with problems – just like Shackleton and his crew or Nansen and Johansen.

These autobiographical reports point to some inescapable conclusions. First, space agencies select resilient people who are good at solving problems and getting along with others (◘ Table 2.2). Second, for most astronauts, spaceflight is their peak life experience. Third, among postflight changes, astronauts consider

◘ **Table 2.2** Astronaut characteristics for long-duration missions
High motivation to achieve
High sense of adventure
Low susceptibility to anxiety
Aged older than 30 years
Emotionally stable
Few symptoms of depression
Low neuroticism
Introverted but socially adept
Not greatly extraverted or assertive
No great need for social interaction
Low demands for social support
Sensitive to needs of others
High tolerance of little mental stimulation
Does not become bored easily
High tolerance to lack of achievement

2

themselves to be changed for the better. These findings in no way detract from the importance of anticipating problems and preparing countermeasures for the challenges of long-duration missions – but equally, they underline the importance of also considering the possibly unique benefits of this great adventure, to the astronauts themselves and to humankind.

Immune System

Immune dysregulation was first observed in astronauts in the 1960s and 1970s, including the Apollo crews, half of whom suffered bacterial infections. While there is much data about astronauts' immune system responses following spaceflight, less is known about what happens *during* a mission (◘ Fig. 2.21). Of the research conducted postflight, data has revealed several changes, including changes in leukocytes, cytokine production, reduced natural killer cell activity, and altered immune responses. Many of these altered immune system changes are related to very high levels of physical and psychological stress that astronauts must endure during their missions [33]. Isolation, confinement, and changed circadian rhythms are all factors that are implicated in the altered immune responses observed in astronauts following long-duration missions [34].

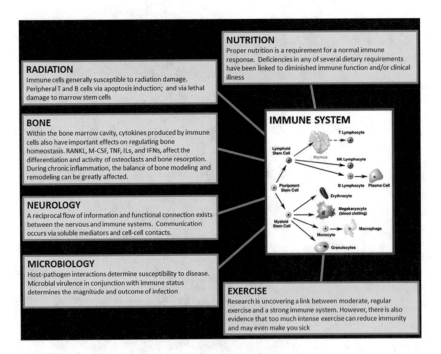

◘ **Fig. 2.21**   How spaceflight affects the immune system. Credit: NASA

Another major factor that has a profound negative impact on immune system function is the exposure to ionizing radiation, although the exact mechanisms that cause radiation-induced immune dysregulation have yet to be fully elucidated. In addition to radiation exposure, the effect of weightlessness causes significant immune system changes, since the absence of gravity alters signaling pathways that are key to T-cell activation.

Radiation

Of all the physiological challenges astronauts must deal with, radiation is the most damaging. Part of the reason is that the body cannot adapt to radiation: the longer an astronaut is in space, the more radiation they will be exposed to. And as we shall see, too much radiation can have significant negative effects on human physiology. But first, a primer.

Galactic Cosmic Rays

Why all the fuss over this particular type of radiation? GCR is unique for the simple reason that it is nigh on impossible to shield against, and here's why. GCRs are highly energetic charged particles that originate outside the Solar System, bulleting along at close to the speed of light, propelled through space by the force of exploding stars [35]. These particles have tremendous energy. The mass of some of these particles, such as iron, combined with their phenomenal speed, enable them to slice through the walls of a spacecraft like the proverbial hot knife through butter.

Astronauts exposed to too much cosmic radiation are at higher risk of developing cancer. And not just any cancer. Dr. Francis Cucinotta [36], one of the world's leading experts on the physiological effects of space radiation, says, "The type of tumors that cosmic ray ions make are more aggressive than what we get from other radiation."

GCRs comprise mainly of protons and alpha particles, which make up about 99 percent of cosmic rays, with the remaining 1 percent composed of heavy nuclei such as lithium, beryllium, and boron. GCRs composed of charged nuclei that are heavier than helium are termed HZE ions. HZE ions are scarce, but because these particles are so highly charged and so heavy, they contribute to a large proportion of an astronaut's radiation dose.

Measuring Galactic Cosmic Rays

The ISS is equipped with a suite of active radiation monitors that provide ground controllers and crew with cumulative exposures and dose rates. We'll discuss these in more detail in ▶ Chap. 7, so what follows is a primer. One example of the monitors carried on ISS is a spectrometer, which provides time-resolved measurements of GCR and the components of GCR such as neutrons, measured via neutron

2

spectroscopy. Why are neutrons so important? Results from various studies over the years have revealed that secondary neutrons contribute up to 30 percent of the total radiation dose that astronauts are exposed to during their tour of duty. Charged particle monitoring is also important because this data provides information about the direction-dependent distribution of charged particles inside the ISS. This data can then be used to calculate accurate radiation data for organ exposure and the consequent risk of that exposure to the particular organ. All in all, the accumulation of this data is collected for the purpose of reducing ambiguity when calculating risk and decreasing the uncertainty when characterizing the radiation environment inside the ISS.

Solar Particle Events

The next component of space radiation is solar particle events (SPEs) (■ Fig. 2.22). SPEs comprise energetic electrons, protons, and alpha particles that are accelerated to speeds approaching light speed by shockwaves that precede coronal mass ejections (CMEs), which occur near solar flare sites. The most energetic SPEs may arrive in LEO within 20 or 30 minutes of the event and may cause significant effects in the Earth's atmosphere. Fortunately, those effects are predictable, to a degree. That's because the Sun's activity follows an 11-year cycle that comprises 4 inactive years (solar minimum) and 7 active years (solar maximum).

During solar maximum, the Sun generates about three CMEs per day, compared to one every 5 days during solar minimum. These events and the timing of these events are significant because a typical CME contains billions of tons of matter, and the shockwaves associated with CMEs may generate magnetic storms that may impact those residing in LEO. During a large SPE, the fluence of protons may

■ **Fig. 2.22** A solar particle event. Credit: NASA

exceed tens of millions of electron volts (MeV) and may significantly increase the radiation dose for crews in LEO. Needless to say, such energies impose significant operational constraints upon mission planners. Unfortunately, there are no ways to reliably predict when an SPE may strike; the best scientists can do is to study the relationship between SPE intensity and the parameters of shock and plasma, to better understand the conditions when these events occur.

Measuring Solar Particle Radiation

Thanks to measurements by the Geostationary Operational Environmental Satellite (GOES) system, scientists have been able to classify the size of SPEs. Most SPEs are classified as A, B, C, M, or X, with each class having a peak flux ten times greater than the previous one. A linear scale exists within each class, which means that a flare rated X2 is twice as powerful as a flare rated X1 (or four times as powerful as one rated M5). The flares that generally are the most disruptive are those in the M and X categories.

To estimate radiation exposure, scientists use the spectra of SPEs to estimate particle intensities. This information is then used to model radiation exposures, but due to the unpredictable characteristics of particle acceleration and propagation of these particles in interplanetary space [37], the spectra of many SPEs are difficult to determine. To partially overcome this problem, scientists use algorithms that sequentially interpolate measured data points. This information may also be used to model radiation exposure. To determine acute radiation risks caused by SPE exposure, a projection code was developed by NASA. Known as the Acute Radiation Risk/BRYNTRRN Organ Dose (ARRBOD) projection code, this software can be used to analyze SPEs. This information, combined with information obtained from human phantom models and dosimetry (see ▶ Chap. 7), can be used to estimate resultant organ doses in those astronauts who encounter a SPE event. It in turn can be used to model the severity of acute radiation sickness and the consequent effects that may include vomiting, nausea, and weakness.

Radiation Dose on Board the International Space Station

The ISS orbits at an altitude of about 400 kilometers (at perigee, it is at 400 km, and at apogee, it orbits at 408 km), which is an altitude above the Earth's primary atmosphere. This means that astronauts are exposed to high fluxes of ionizing radiation, the primary sources of which are GCRs. In addition to GCRs, ISS crews are exposed to particles trapped in the Van Allen belts and SPEs [38, 39, 40].

Another key element in calculating the radiation exposure of astronauts on board the orbiting outpost is the orbital inclination of the ISS, which is 52°. This inclination means that the station passes through the South Atlantic Anomaly (SAA) every day. The SAA, which is located east of Argentina, is characterized by an anomaly in the Earth's geomagnetic field which results in energetic particles

◘ Fig. 2.23 Genetic material that is bombarded by radiation may be difficult to repair. Credit: NASA

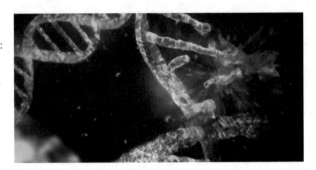

penetrating lower altitudes than normal. This means that when the ISS passes through the SAA, astronauts are exposed to higher levels of ionizing radiation. Astronauts are already at risk of developing cancer simply by traveling to space. Astronauts on the ISS receive 80 millisieverts (mSv) during a six-month increment, whereas people on Earth receive 2 mSv a year. Another comparative metric is the chest radiograph dose, which is 0.02µSv per hour. And if you happen to spend a lot of time flying commercial, the radiation dose per hour is about 0.3 to 5.7µSv. Spend too much time on orbit and astronauts hit their career radiation limit, which equates to an increase of 3 percent risk of developing cancer in their lifetime. Any astronaut heading for Mars will be exposed to between 1 and **5 Sv** and therefore be guaranteed to hit – and exceed – that limit by virtue of being exposed to killer radiation for the best part of 2 years. But remember, this is not just any type of radiation. This radiation isn't stopped by shielding, and there is a lot of it. Out there in deep space, it would only take about 3 days for every cell in every crew-member to be hit by a high-energy proton. Now, for some cells, it isn't much of a problem, but when these careening nuclei hit important cells such as DNA, mutations may result (◘ Fig. 2.23).

Radiation Risk

In the United States, the incidence rate of cancer is 38.5% (according to the National Cancer Institute – ▶ www.cancer.gov – based on statistics between 2008–2012). If you exposed 100 people (which is the capacity of SpaceX's Mars-bound Starship incidentally) to the 1 Sv of radiation that Mars astronauts will be exposed to, 61 of them will be diagnosed with cancer. By virtue of the unique characteristics of GCR, these cancers would typically be lung, breast and colorectal cancers, meaning half these astronauts would die. Scientists have modeled the dangers of GCRs during a manned Mars mission and have calculated that exposure to radiation on such a trip would shorten an astronaut's lifespan by between 15 and 24 years.

Radiation and Crew Health

It sounds like a lot of doom and gloom, which is why experts like Dr. Cucinotta [36] recommend that space agencies gather a lot more data about how GCR affects crew health and how these effects may be mitigated. For example, some of the immune responses caused by GCR exposure are similar to those in inflammatory diseases. In such cases, oxidants are produced that change intercellular signaling, but it is possible that nonsteroidal anti-inflammatory drugs could be taken by astronauts during their Mars mission to mitigate some of the effects of cancer. Another more serious effect of GCR is that it may accelerate the onset of symptoms similar to those exhibited by Alzheimer's patients [41, 42, 43]:

» Galactic cosmic radiation poses a significant threat to future astronauts,". "The possibility that radiation exposure in space may give rise to health problems such as cancer has long been recognized. However, this study shows for the first time that exposure to radiation levels equivalent to a mission to Mars could produce cognitive problems and speed up changes in the brain that are associated with Alzheimer's disease.

» Professor M. Kerry O'Banion, M.D., Ph.D., University of Rochester Medical Center Department of Neurobiology and Anatomy.

O'Banion's research focused on how GCR affects the central nervous system (CNS) and the news was less than rosy. Much of his research has been conducted at the NASA Space Radiation Laboratory at Brookhaven National Laboratory on Long Island. It's a place that was chosen for its accelerators, which can collide matter at extremely high speeds, thereby replicating what happens in space. O'Banion's research examined the effect of GCR-equivalent radiation on cognitive function (i.e., how long it took mice to find their way through a maze). To do this, mice were exposed to various doses of radiation comparable to levels Mars-bound astronauts will be subjected to. The mice were subject to recall tests, and researchers found that mice exposed to radiation were more likely to fail the tests (they couldn't find their way through the maze), a finding that indicated cognitive impairment [41].

» Because iron particles pack a bigger wallop it is extremely difficult from an engineering perspective to effectively shield against them. One would have to essentially wrap a spacecraft in a six-foot block of lead or concrete.

» Professor M. Kerry O'Banion, M.D., Ph.D., University of Rochester Medical Center Department of Neurobiology and Anatomy.

After examining the mice more closely, researchers found that the brains of the mice exhibited signs of vascular alteration and larger than normal amount of beta-amyloid, which happens to be a signature of the Alzheimer's. The researchers concluded that exposing astronauts to GCR for an extended duration might accelerate Alzheimer's. So what can astronauts do? Well, they can shield themselves from the radiation, a subject discussed in ▶ Chap. 7, and they can monitor their exposure to radiation while they are inside the vehicle.

Intravehicular Radiation

On board the ISS, passive dosimeters are located in each pressurized module [38, 44]. These dosimeters measure time-integrated absorbed doses, which change according to the station's altitude and position in the solar cycle. The requirements of these dosimeters are defined in the International Space Station Medical Operations Requirements Document (ISS MORD SSP 50260), a document that states the dose limits and radiation exposure practices across all mission phases. These radiation monitors must perform the following functions:
1. Measure cumulative radiation dose.
2. Downlink linear energy transfer data.
3. Provide direction-dependent distribution radiation data for inside and outside the ISS.
4. Downlink data for frequent analysis.
5. Downlink dose rate from charged particle monitoring equipment.
6. Alert the crew when exposure rates exceed set threshold.

Biomedical Consequences of Exposure to Space Radiation

The biological effects of exposure to space radiation can be divided into *acute* and *chronic*. Acute effects are the result of exposure to high radiation doses, which may be caused by SPEs, whereas chronic effects are caused by extended exposure to space radiation. Potential effects of either type of exposure include *direct* and *indirect* damage to genetic material, biochemical alterations of cells and/or tissues, carcinogenesis, degenerative tissue effects, and cataracts. The extent of these effects is determined by the type of radiation, its flux, and the energy spectrum, factors that are incompletely understood. Other factors that determine radiation damage include age at exposure, gender, and susceptibility to radiation.

The quantitative physiological effects of radiation are also poorly understood, due partly to misinterpretation of the exact mechanisms and processes that concern DNA repair. For example, experiments in the 2010s revealed a number of uncertainties that apply to the quality factors used in radiation protection; some studies indicated that physiological damage caused by a specific exposure was only half that previously estimated, a finding explained by the fact that low energy protons inflict more damage than high energy ones. Why? Very simply, because low energy protons take longer to pass through the body and therefore have more time to interact with the tissues.

Yet another poorly understood mechanism is that of relative biological effectiveness (RBE), which is determined to a large degree by radiation type and kinetic

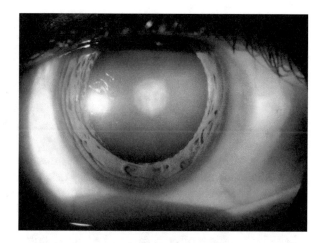

Fig. 2.24 Cataracts are yet another risk of spending time in deep space. Credit: Rakesh Ahuja, MD

energy. How RBE correlates with tumor type or cancer progression is practically unknown because most of the limited experimental data has been conducted on mice, and it is difficult to extrapolate and apply mice data to humans. It is even more of a challenge to use that data to estimate health risks for cancer, cataracts (■ Fig. 2.24), and CNS risks. To gain even a cursory insight into the problem will require many, many, *many* more astronauts conducting 1-year (or longer) increments followed by post-mission observation times of at least 10 years. At least. Given the ISS is due to be retired in 2028, this goal is impossible. And even it was, extrapolating data from the ISS to deep space is extremely limited at best in terms of making accurate risk predictions for those who eventually venture beyond Earth orbit (BEO). In short, there are myriad knowledge gaps regarding the potential acute and late biomedical risks from GCRs and SPEs, but what follows is some of what we know.

Central Nervous System Effects

The potential acute and late risks to the CNS from GCRs and SPEs have not been a major consideration for crews on board the ISS because these astronauts are exposed to relatively low-to-moderate doses of ionizing radiation compared to deep space doses.

» I'm having these light flashes. I'm seeing this, like, light flashing in my eyeballs. It was like fireworks in your eyeballs. It was spectacular.

» Charles Duke, Lunar Module Pilot, Apollo 16

The risk presented by GCRs was evident during the Apollo era [45] when astronauts reported the "light flash" phenomenon caused by HZE nuclei traversing through the retina. As these nuclei traverse, they cause a microlesion. In addition to microlesions in the eye, exposure to HZE nuclei, at doses similar to ones that

2

◻ **Table 2.3** MSL measurements for average dose rate on cruise phase to Mars and on Mars surface [4]	
	GCR dose equivalent rate (mSv/day)
RAD cruise to Mars	1.84 ± 0.33
RAD Mars surface	0.70 ± 0.17

astronauts will be exposed to during a Mars mission, causes neurocognitive deficits, such as operant reactions. Other CNS risks include detriments in short-term memory and altered motor function. These are discussed briefly here.

Radiation protection for astronauts is based on a risk of a 1 in 33 probability of death by cancer caused by occupational exposure, a limit that compares with a 1 in 270 risk of loss of crew caused by a flight failure. This risk estimate is determined by human epidemiology combined with quality factors, risk projection models, and experimental models, but this method cannot be applied to estimate CNS risks in deep space [46]. The reason for this is that there have been very few humans that have ventured beyond Earth orbit, which means human scaling is next to impossible. What is known is that GCR comprises protons, helium nuclei, and nuclei that travel with high charge and great energy – also known as HZE nuclei. The energy of these nuclei may range from tens of millions of electron volts (MeV)/u to more than 10,000 MeV/u[2]. But these energies do not tell the whole story, because secondary particles are generated as the nuclei pass through shielding and tissue. And since GCR nuclei have such high energies, they are capable of passing through hundreds of centimeters of material. To get a better understanding of the GCR environment on a trip to Mars, NASA flew the RAD on the MSL (◻ Table 2.3)[1,2].

◻ Table 2.3 shows the GCR dose equivalent measured by the RAD/MSL study. By comparison, astronauts spending 6 months on the ISS are exposed to 80 mSv or less than a quarter of the exposure of astronauts spending more than 6 months on a trip to Mars. The reason for the difference is that astronauts in LEO are partly protected by the Earth's magnetic field, which repels GCR nuclei with energies below 1000 MeV/u. With Mars astronauts taking such a radiation hit, it will be necessary to find out as much as possible about the biomedical effects of exposure to GCR. One option is to test nonhuman primates (NHP), since NHPs and humans share a number of physiological and neurobiological characteristics. For example, NHPs are used to investigate infectious diseases, Alzheimer's, and

---

1   LET is the retarding force acting on a charged ionizing particle as it passes through material, whether that material happens to be a spacecraft or an astronaut. The term describes how much energy the particle transfers to the material traversed per unit distance. LET also depends on the type of radiation and the material traversed.

2   MeV is short for *megaelectron volt* and is equivalent to 1 million electron volts (eV). 1 eV is the amount of kinetic energy gained by an electron as it accelerates through an electric potential difference of 1 volt.

strokes. Rodents can also be used, although the number of cross-species differences makes clinical determination of CNS health risks. For example, many of the cognitive deficits are known to originate in the frontal cortex in humans, but this area is underdeveloped in rodents. Another example is the difference in risk assessment; death for 50% of a population from a fixed level of radiation exposure occurs at a lower level of exposure for humans than it does for certain strains of rodents. But rodents are the more favored test subject when it comes to radiation research because NHP studies require much higher costs and more thorough ethical review.

Behavioral Studies of CNS Risks

One radiation-related topic that has received media attention in recent years is the suggestion that exposure to deep space radiation may cause cognitive deficits. The aforementioned study that investigated this [41] exposed rats to 1000 MeV/u before testing their spatial memory in a radial maze. In this study, the exposed rats committed more errors than control rats and were unable to develop a spatial strategy to make their way through the maze. A similar study [8] examined mice that had been subject to 2 weeks of whole-body irradiation. The results revealed impaired novel object recognition and reduced spatial memory. Another study exposed Wistar rats to 1000 MeV/u and tested the rats 3 months after exposure. The test in this study was an attentional set shifting task (AST)[3] in which only 17% of the irradiated rats were able to complete compared with 78% of control rats.

Altered Neurogenesis[4]

Research has revealed that neurogenesis may be sensitive to radiation [46, 47] which in turn may result in cognitive deficits such as memory [41]. Furthermore, studies have indicated not only that exposure to high doses of radiation may inhibit the generation of neuronal progenitor cells but that those cells that are generated may not be fully functional [42]. These studies were conducted on mice using doses of 1000 MeV/u, which provide some insight into the mechanism of radiation-induced cognitive injury, but for reasons indicated earlier, scaling these results to humans is limited.

---

3   The AST measures attention and cognitive flexibility in rats. It is based on the intradimensional/extradimensional component of the Cambridge Neuropsychological Test Automated Battery (CANTAB) which is used to assess cognitive dysfunction in humans. An attentional set is created when a person learns that a set of rules can be used to distinguish between relevant and irrelevant cues

4   Neurogenesis is a term that describes the formation of neurons. In adults, this process occurs in the subventricular zone (SVZ) and the subgranular zone of the brain. Scientists are still researching the role that neurogenesis plays in cognition. Since the formation of neurons may be sensitive to radiation, it is possible that long-term exposure may result in cognitive deficits.

Oxidative Damage[5]

Oxidative stress is thought to be implicated in Alzheimer's disease, heart failure, and chronic fatigue syndrome. Since radiation has been shown to increase oxidative damage, oxidative stress represents yet another mechanism of radiation-induced cognitive injury. Since antioxidants prevent such damage under normal conditions, it would seem logical to suggest that astronauts eat food that contains high levels of antioxidants. For example, a diet high in blueberries should help offset oxidative stress. Or melatonin perhaps? Melatonin has high antioxidant properties and studies have shown that it inhibits neurogenesis. The problem with research to date is that studies have used high-dose rates and the biological effects of radiation are different at low-dose rates. Furthermore, studies that have investigated the supposed beneficial effects of antioxidants found no evidence that supplementation actually works. In fact, some studies revealed that antioxidants such as vitamin A, vitamin E, and β-carotene might actually be more damaging because taking extra amounts of antioxidants would help the body rescue cells that had been damaged by radiation and that this might alter DNA repair.

» This study shows for the first time that exposure to radiation levels equivalent to a mission to Mars could produce cognitive problems and speed up changes in the brain that are associated with Alzheimer's disease. These findings clearly suggest that exposure to radiation in space has the potential to accelerate the development of Alzheimer's disease. This is yet another factor that NASA, which is clearly concerned about the health risks to its astronauts, will need to take into account as it plans future missions.

» Dr. Kerry O'Banion, University of Rochester Medical Center.

Alzheimer's Disease

Alzheimer's disease is a neurodegenerative disease that causes dementia in most cases. Common symptoms include short-term memory loss, language problems, disorientation, lack of motivation, and behavioral problems. The disease, which is chronic, begins slowly and symptoms become worse with time. The cause of the disease is not completely understood but the majority of the risk is believed to be genetic. There are no treatments that can stop the disease or even slow its progression. It has been shown in mice that exposure to radiation accelerates the onset of age-related neuronal dysfunction that results in symptoms similar to those exhibited by those suffering from Alzheimer's [41, 43].

---

5   Oxidative stress is a term that describes the imbalance between the production of free radicals and the ability of the body to neutralize these free radicals through the use of antioxidants. Free radicals are molecules that contain oxygen. These molecules have one or more unpaired electrons, which means they are very reactive with other molecules which in turn means they are capable of chemically interacting with and destabilizing cells such as DNA. Under normal conditions, antioxidants prevent these reactions.

Radiation-Induced Bone Loss

Another process affected by radiation is bone remodeling. In zero gravity or reduced gravity, bone remodeling is disrupted, resulting in a loss of BMD. Astronauts on board the ISS typically lose between 1.0 and 1.2 percent of BMD per month. This equates to an overall loss of more than 7 percent during a typical 6-month increment, which in turn results in a two to threefold increase in fracture risk. For astronauts in LEO, this rate of BMD loss is fairly predictable, but beyond LEO, the radiation environment is much harsher, and the rate of BMD loss less predictable. This is because bone is damaged more by higher doses of radiation, a fact long documented by the persistent decline in bone volume following exposure to therapeutic radiation in cancer patients [48, 49].

Compounding the effect of osteoradionecrosis is the effect that radiation has on fracture sites [49]. To better understand this, it is necessary to familiarize ourselves with the biological damage caused by radiation. There are two types of radiation: *non-ionizing* and *ionizing*. Non-ionizing radiation does not cause significant biological damage because this type of radiation does not displace electrons from an atom. Ionizing radiation on the other hand has sufficient energy to displace electrons from an atom, thereby creating an ionization event. The energy of this ionization event can break molecular bonds and thereby cause biological damage such as single-strand or double-strand breaks in DNA. While the body is tremendously resilient in its capacity to repair radiation damage, some cells ultimately die in this onslaught of ionizing radiation. Worse, some cells may actually propagate the ionized-induced damage to progeny.

We already know that the space radiation environment comprises a mix of ions generated by SPEs and GCR. We also know, thanks to the data sent back from the RAD that was strapped onto Curiosity, that the transit to Mars will result in a radiation exposure of more than 1 Sievert. This means that tissue dose rates from space radiation will be about 1–2.5 mSv per day (the *annual* terrestrial dose incidentally), but solar flare dose rates may increase this number to more than 100 mSv/day, even if inside a shielded vehicle. Furthermore, an astronaut conducting a deep space EVA during a solar flare event may be exposed to a dose rates as high as 250 mSv per day. By comparison, cancer patients receive daily dose (fractions) of about 6 Sv targeted at the tumor, but these doses are delivered over a period of minutes [48].

We have a fairly good understanding of these processes because ionizing radiation has long been used as a treatment for malignancies and has been a factor in reducing cancer mortality. One of the main reasons BMD is reduced following irradiation is because osteoblasts and osteoclasts are damaged (a quick review: osteoblasts and osteoclasts are two bone cells that work together to remodel bone; the osteoclasts break down bone, and the osteoblasts build up bone). But when these bone cells are damaged, bone formation is impaired due to cell cycle arrest. One of the processes by which the osteoclasts and osteoblasts are damaged is by oxidative stress caused by the radiation since it is this oxidative stress that damages osteoprogenitors. To begin with, irradiation causes an increase in osteoclast number which thereby causes osteoporosis. Shortly after exposure, there is a decline in the number of osteoclasts and osteoblasts, which results in suppression of bone remodeling and degradation of bone quality.

2

The side effects of radiation treatment have concerned oncologists and will be of concern to flight surgeons responsible for the health of astronauts embarked on exploration class missions (ECM) beyond LEO. This is because bones within the irradiated area are at a much higher fracture risk. For example, patients undergoing breast cancer treatment may have rib fracture rates that exceed 15 percent. This is of concern to astronauts on long duration because their bones will already be weakened due to the loss of BMD simply as a consequence of being in microgravity. Consequently, these astronauts may be at high risk of traumatic and/or spontaneous fracture.

So, the effect of radiation results in a cascade of changes. In addition to the reduction in bone mineral density and the impact on healing, the way in which bone is weakened will be of concern to flight surgeons tasked with keeping ECM crewmembers fracture-free. For example, the loss of trabecular bone means cortical bone must now deal with a greater proportion of loads on the skeleton. This in turn means that cortical bone will be increasingly less able to resist the torsional and bending loads, a change that may be exacerbated by any defect in the bone such as a porous hole. The net effect of all these radiation-induced effects is an overall disruption of load distribution that results in a compromised structural integrity of the bone. For astronauts about to land on Mars, this is not an optimum situation. It is important to remember that after their ISS increments, astronauts return to Earth with increased fracture risk, but this doesn't mean a fracture is imminent. It just means that due to the loss of BMD, there is a greater chance of fracture after their return. But this increased fracture risk is dramatically reduced, thanks to the rehabilitation schedule that astronauts follow after return to Earth, which results in regeneration of bone.

Radiation-induced bone loss is termed *osteoradionecrosis*. *Osteitis* is a condition in which the bone's ability to withstand trauma is reduced. In this condition, nonhealing bone may be susceptible to infection, and the ability of the bone to heal is further complicated by hypovascularization. Basically, as the body is subjected to more and more radiation, the very small blood vessels inside the bone are destroyed. This is devastating because these blood vessels carry nutrients and oxygen to the bone. Without blood vessels to do this, the bone simply dies. On Earth, one treatment option for patients with osteoradionecrosis is hyperbaric oxygen therapy (although even with this treatment, less than 30 percent of patients will survive), but this will not be an option on an interplanetary spaceship.

Key Terms
- Advanced Resistive Exercise Device (ARED)
- Beyond Earth Orbit (BEO)
- Central Nervous System (CNS)
- Cerebrospinal Fluid (CSF)
- Coronal Mass Ejection (CME)
- Deoxyribonucleic Acid (DNA)
- Exploration Class Mission (ECM)

- Food Intake Tracking (FIT)
- Galactic Cosmic Radiation (GCR)
- Isolated Confined Extreme (ICE)
- International Space Station (ISS)
- Intracranial Pressure (ICP)
- Intraocular Pressure (IOP)
- Low Earth Orbit (LEO)
- Linear Energy Transfer (LET)
- Mars Science Laboratory (MSL)
- Magnetic Resonance Angiography (MRA)
- Magnetic Resonance Imaging (MRI)
- Magnetic Resonance Venogram (MRV)
- Optical Diameter (OD)
- Optic Nerve Sheath Diameter (ONSD)
- Optical Coherence Tomography (OCT)
- Radiation Assessment Detector (RAD)
- Relative Biological Effectiveness (RBE)
- South Atlantic Anomaly (SAA)
- Solar Particle Event (SPE)
- Vision Impairment Intracranial Pressure (VIIP)

❓ Review Questions

1. How much calcium is excreted by the body per day in microgravity?
2. What is the function of the osteoblasts in bone remodeling?
3. What is the function of slow twitch muscle fibers?
4. What is meant by the term hypertrophy?
5. Explain why some astronauts suffer motion sickness.
6. How is ICP implicated in the VIIP syndrome?
7. Where would you expect to find the laminar cribrosa sclera?
8. What is meant by the term *diuresis*?
9. What is meant by the term *salutogenesis*?
10. What happens to natural cell activity in space?
11. What is the South Atlantic Anomaly?
12. How does radiation affect the CNS?

References

1. Stein, T. P. (2013). Weight, muscle and bone loss during space flight: another perspective. *European Journal of Applied Physiology, 113*, 2171–2181.
2. Orwoll, E. S., et al. (2013). Skeletal health in long-duration astronauts: nature, assessment, and management recommendations from the NASA bone summit. *Journal of Bone and Mineral Research, 28*, 1243–1255.
3. Lang, T., LeBlanc, A., Evans, H., et al. (2006). Cortical and trabecular bone mineral loss from the spine and hip in long-duration spaceflight. *Journal of Bone and Mineral Research, 19*, 1006–1012.

2

4. Cann, C. (1997). Response of the skeletal system to spaceflight. In S. E. Churchill (Ed.), *Fundamentals of space life sciences* (Vol. 1, pp. 83–103). Malabar: Krieger publishing company.
5. Buckey, J. C. (2006). Bone loss: managing calcium and bone loss in space. In M. R. Barratt & S. L. Pool (Eds.), *Space physiology* (pp. 5–21). New York: Oxford University Press.
6. Whitson, P. A., Pietrzyk, R. A., & Sams, C. F. (2001). Urine volume and its effects on renal stone risk in astronauts. *Aviation, Space, and Environmental Medicine, 72*, 368–372.
7. Whitson, P. A., Pietrzyk, R. A., Morukov, B. V., & Sams, C. F. (2001). The risk of renal stone formation during and after long duration space flight. *Nephron, 89*, 264–270.
8. Loehr, J. A., et al. (2011). Musculoskeletal adaptations to training with the advanced resistive exercise device. *Medicine and Science in Sports and Exercise, 43*, 146–156.
9. Schneider, S. M., et al. (2003). Training with the international space station interim resistive exercise device. *Medicine and Science in Sports and Exercise, 35*, 1935–1945.
10. Loehr, J. A., et al. (2015). Physical training for long-duration spaceflight. *Aerospace Medicine and Human Performance, 86*, A14–A23.
11. Heer, M., & Paloski, W. H. (2006). Space motion sickness: incidence, etiology, and countermeasures. *Autonomic Neuroscience, 129*, 77–9. Epub 2006 Aug. 28.
12. Zhang, L. F., & Hargens, A. R. (2014). Intraocular/Intracranial pressure mismatch hypothesis for visual impairment syndrome in space. *Aviation, Space, and Environmental Medicine, 85*, 78–80.
13. Mader, T. H., Gibson, C. R., Pass, A. F., Kramer, L. A., Lee, A. G., Fogarty, J., Tarver, W. J., Dervay, J. P., Hamilton, D. R., Sargsyan, A., Phillips, J. L., Tran, D., Lipsky, W., Choi, J., Stern, C., Kuyumjian, R., & Polk, J. D. (2011). Optic disc edema, globe flattening, choroidal folds, and hyperopic shifts observed in Astronauts after long-duration space flight. *Ophthalmology, 118*(10), 2058–2069.
14. Alexander, D. J., Gibson, C. R., Hamilton, D. R., Lee, S. M., Mader, T. H., Otto, C., Oubre, C. M., Pass, A. F., Platts, S. H., Scott, J. M., Smith, S. M., Stenger, M. B., Westby, C. M., & Zanello, S. B. *Human research program human health countermeasures element evidence report.* Risk of spaceflight-induced intracranial hypertension and vision alterations. July 12, 2012. Version 1.0
15. Mader, T. H., et al. (2011). Optic disc edema, globe flattening, choroidal folds, and hyperopic shifts observed in astronauts after long-duration space flight. *Ophthalmology, 118*, 2058–2069.
16. Mader, T. H., et al. (2013). Optic disc edema in an astronaut after repeat long-duration space flight. *Journal of Neuro-Ophthalmology, 33*, 249–255.
17. Alperin, N., Hushek, S. G., Lee, S. H., Sivaramakrishnan, A., & Lichtor, T. (2005). MRI study of cerebral blood flow and CSF flow dynamics in an upright posture: the effect of posture on the intracranial compliance and pressure. *Acta Neurochirurgica. Supplement, 95*, 177–181.
18. Nelson, E. S., Mulugeta, L., & Myers, J. G. (2014). Microgravity-Induced Fluid Shift and Ophthalmic Changes. *Life, 4*(4), 621–665.
19. Kramer, L. A., Sargsyan, A. E., Hasan, K. M., Polk, J. D., & Hamilton, D. R. (2012). Orbital and intracranial effects of microgravity: Findings at 3-T MR imaging. *Radiology, 263*, 819–827.
20. Jonas, J. B., Berenshtein, E., & Holbach, L. (2003 Dec). Anatomic relationship between lamina cribrosa, intraocular space, and cerebrospinal fluid space. *Investigative Ophthalmology & Visual Science, 44*(12), 5189–5195.
21. Friedman, D. I. (2007). Idiopathic intracranial hypertension. *Current Pain and Headache Reports, 11*(1), 62–68.
22. Hargens, A. R., & Richardson, S. (2009). Cardiovascular adaptations, fluid shifts, and countermeasures related to space flight. *Respiratory Physiology & Neurobiology, 169*, S30–S33.
23. Convertino, V. A., & Cooke, W. H. (2007). Vascular functions in humans following cardiovascular adaptations to spaceflight. *Acta Astronautica, 60*, 259–266.
24. Norsk, P. (2005). Cardiovascular and fluid volume control in humans in space. *Current Pharmaceutical Biotechnology, 6*, 325–330.
25. Kirsch, K. A., Baartz, F. J., Gunga, H. C., & Rocker, L. (1993). Fluid shifts into and out of superficial tissues under microgravity and terrestrial conditions. *The Clinical Investigator, 71*, 687–689.
26. Foldager, N., et al. (1996). Central venous pressure in humans during microgravity. *Journal of Applied Physiology, 81*, 408–412.

27. Norsk, P. (2014). Blood pressure regulation IV: adaptive responses to weightlessness. *European Journal of Applied Physiology, 114*, 481–497.

28. Baevsky, R. M., et al. (2007). Autonomic cardiovascular and respiratory control during prolonged spaceflights aboard the International Space Station. *Journal of Applied Physiology, 103*, 156–161.

29. Verheyden, B., et al. (2008). Cardiovascular control in space and on earth: the challenge of gravity. *IRBM, 29*, 287–288.

30. Musson, D. M., & Helmreich, R. L. (2005). Long-term personality data collection in support of spaceflight and analogue research. *Aviation, Space, and Environmental Medicine, 76*(Suppl), B119–B125.

31. Collins, D. L. (2003). Psychological issues relevant to astronaut selection for long-duration space flight: a review of the literature. *Journal of Human Performance in Extreme Environments, 7*, 43–67.

32. Dion, K. L. (2004). Interpersonal and group processes in long-term spaceflight crews: perspectives from social and organizational psychology. *Aviation, Space, and Environmental Medicine, 75*(Suppl 7), C36–C43.

33. Mehta, S. K., Crucian, B., Pierson, D. L., et al. (2007). Monitoring immune system function and reactivation of latent viruses in the Artificial Gravity Pilot Study. *Journal of Gravitational Physiology, 14*, P21–P25.

34. Gmunder, F. K., Konstantinova, I., Cogoli, A., et al. (1994). Cellular immunity in cosmonauts during long duration spaceflight on board the orbital MIR station. *Aviation, Space, and Environmental Medicine, 65*, 419–423.

35. Getselev, I., Rumin, S., Sobolevsky, N., Ufimtsev, M., & Podzolko, M. (2004). Absorbed dose of secondary neutrons from galactic cosmic rays inside the international space station. *Advances in Space Research, 34*, 1429–1432.

36. Cucinotta, F. A., Kim, M. Y. & Chappell, L. (2013). Space radiation cancer risk projections and uncertainties-2012. NASA TP 2013–217375..

37. Zeitlin, C., Hassler, D. M., Cucinotta, F. A., Ehresmann, B., Wimmer-Schweingruber, R. F., Brinza, D. E., et al. (2013). Measurements of energetic particle radiation in transit to Mars on the Mars science laboratory. *Science, 340*, 1080–1084.

38. Akopova, A. B., Manaseryan, M. M., Melkonyan, A. A., Tatikyan, S. S., & Potapov, Y. (2005). Radiation measurement on the International Space Station. *Radiation Measurements, 39*, 225–228.

39. Kodaira, S., Kawashima, H., Kitamura, H., Kurano, M., Uchihori, Y., Yasuda, N., Ogura, K., Kobayashi, I., Suzuki, A., Koguchi, Y., Akatov, Y. A., Shurshakov, V. A., Tolochek, R. V., Krasheninnikova, T. K., Ukraintsev, A. D., Gureeva, E. A., Kuznetsov, V. N., & Benton, E. R. (2013). Analysis of radiation dose variations measured by passive dosimeters onboard the International Space Station during the solar quiet period (2007–2008). *Radiation Measurements, 49*, 95–102.

40. Cucinotta, F. A., Kim, M. Y., Willingham, V., & George, K. A. (2008). Physical and Biological Organ Dosimetry Analysis for International Space Station Astronauts. *Radiation Research, 170*(1), 127–138.

41. Heneka, M. T., & O'Banion, M. K. (2007). Review article Inflammatory processes in Alzheimer's disease. *Journal of Neuroimmunology, 184*, 69–69.

42. Cucinotta, F. A., Alp, M., Sulzman, F. M., & Wang, M. (2014). Space radiation risks to the central nervous system. *Life Sciences in Space Research, 2*, 54–69.

43. Cherry, J. D., Liu, B., Frost, J. L., Lemere, C. A., Williams, J. P., et al. (2012). Galactic cosmic radiation leads to cognitive impairment and increased Aβ plaque accumulation in a mouse model of Alzheimer's disease. *PLoS One, 7*(12), e53275.

44. Dietze, G. et al. (2013). *Assessment of astronaut exposures in space.* ICRP Publication 123. Ed. Clement, C. Annals of the International Commission on Radiological Protection 42.

45. Pinsky, L. S., Osborne, W. Z., Bailey, J. V., Benson, R. E., & Thompson, L. F. (1974). Light flashes observed by astronauts on Apollo 11 through Apollo 17. *Science, 183*(4128), 957–959.

46. Cucinotta, F. A., Wang, H., & Huff, J. L. (2009. *Risk of acute or late central nervous system effects from radiation exposure.* Human Health and Performance Risks of Space Exploration Missions NASA SP 2009, p. 345. Chapter 6.

2

47. Britten, R. A., Davis, L. K., Johnson, A. M., Keeney, S., Siegel, A., Sanford, L. D., Singletary, S. J., & Lonart, G. (2012). Low (20 cGy) doses of 1 GeV/u ^{56}Fe-particle radiation lead to a persistent reduction in the spatial learning ability of rats. *Radiation Research, 177*(2), 146–115.
48. Hopewell, J. W. (2003). Radiation-therapy effects on bone density. *Medical and Pediatric Oncology, 41*, 208–211.
49. Sugimoto, M., Takahashi, S., Toguchida, J., Kotoura, Y., Shibamoto, Y., & Yamamuro, T. (1991). Changes in bone after high-dose irradiation. Biomechanics and histomorphology. *Journal of Bone and Joint Surgery, 73*, 492–497.

Suggested Reading

Nicogossian, A. E., Huntoon, C. L., Pool, S. L., & Johnson, P. C. (1988). *Space Physiology and Medicine*. Philadelphia: Lea and Febiger.
Principles of clinical medicine for space flight. New York: Springer Science and Business Media; 2008.
Survival and Sacrifice in Mars Exploration by Erik Seedhouse. Springer-Praxis.
Endurance by Alfred Lansing.

Open-Loop Vs. Closed-Loop Life Support Systems

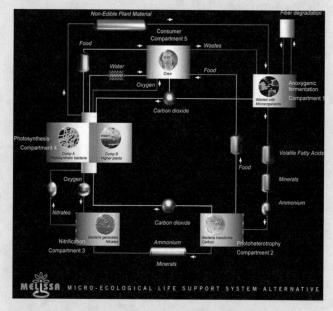

MeLiSSA. An example of a closed-loop life support system concept. Credit: ESA

Contents

© Springer Nature Switzerland AG 2020
E. Seedhouse, *Life Support Systems for Humans in Space*,
https://doi.org/10.1007/978-3-030-52859-1_3

After reading this chapter, you should be able to:
- Explain the difference between an open life support system and a closed life support system
- State the primary life support system functions for food, water, waste, atmosphere, and crew safety
- Explain the difference between fail-safe and fail-operational
- Explain how specific mission factors impact life support system design
- List the primary subsystems of a life support system
- List the life support interfaces for each life support subsystem
- List the radiation exposure value for an astronaut spending 6 months on the ISS
- Explain how human metabolic rate is calculated
- Explain the significance of protein synthesis and inadequate energy intake on muscle mass
- Describe the functions of atmosphere management
- Explain how carbon dioxide is removed from the ISS atmosphere
- Explain the challenges of devising a closed-loop life support system

Introduction

There are various life support system categories – open-loop, closed-loop, regenerative, hybrid, bioregenerative, and physicochemical. This chapter explains the difference between each type by highlighting distinguishing characteristics of each. Many people are surprised that the life support system on board the International Space Station is classified as a partially closed system (93 percent closure) and wonder why it isn't fully closed. Well – as you will begin to understand by reading this book – life support system closure remains an extraordinary engineering challenge. Just consider all the functions a life support system has to perform. And it's not just being able to perform all those functions; a life support system has to perform these functions reliably for months and years with minimal maintenance. Ever wonder what the astronauts do when they're not performing science or exercising? Maintaining the life support system!

Open-Loop Vs. Closed-Loop Life Support Systems

There are two primary types of life support systems (LSS): open-loop life support systems and closed-loop life support systems. We'll start with a general overview and then dive into the details.

In an open-loop life support system, all required human inputs are supplied as consumables. These consumables include food, oxygen, and water. Additionally, none of these consumables are recycled. The advantage of such a system is that it is technically simple. Very simple. Another advantage is that it is reliable. The disadvantages depend on mission length. Those consumables represent a mass penalty, and if you rely on an open-loop LSS, the longer your mission, the greater your mass penalty [1, 2, 3].

3

◘ Table 3.1   Life support functions

Life support category	Functions
Maintaining the atmosphere	Pressure control Temperature and humidity control Atmosphere recycling Ventilation Contamination control
Food	Providing food [Producing food: for long-duration missions]
Water management	Providing potable water Recovering water Providing water for hygiene Processing wastewater
Waste management	Collecting waste Storing waste Processing waste
Crew safety	Fire detection and suppression Radiation shielding [For long-duration missions: exercise countermeasures]

Let us take oxygen. Each crewmember requires 840 grams of oxygen per day. That's 3.3 kilograms per day for a crew of four. Multiply that by 1000 days, which is about how long a return trip to Mars might take, and you have more than 3000 kilograms. That's a lot of oxygen that has to be carried on your spacecraft. And that does not include an amount for contingencies.

So, the obvious solution for long-duration missions is to opt for a closed LSS. The problem is no such system exists except here on Earth. On Earth, we have a fully closed LSS in which biological and chemical processes act to recycle wastes into life support resources. Another classification of life support systems is to group them into *non-regenerative* (open-loop) and *regenerative* (closed-loop). And a third type of classification is to use the following terms: *physicochemical life support system* for life support systems that provide some regenerative functions using physicochemical processes and *hybrid life support system* for regenerative life support systems that utilize physicochemical *and* biological processes. Regardless of which type of system, each must fulfil the functions listed in ◘ Table 3.1.

Life Support System Design Factors

Given that closed life support systems, whether they be physicochemical, regenerative, or hybrid, seem to be the preferred system for long-duration flights (◘ Figs. 3.1, 3.2, 3.3 and 3.4), surely space agencies should adopt these as the life support system of choice. But it isn't that simple. Despite working for decades, the very best life

Fig. 3.1 Life support features as a function of mission length: 1–12 hours. Credit: NASA

Fig. 3.2 Life support features as a function of mission length: 1–7 days. Credit: NASA

3

 Fig. 3.3   Life support features as a function of mission length: 12 days–3 months. Credit: NASA

 Fig. 3.4   Life support features as a function of mission length: 3 months–3 years. Credit: NASA

support scientists and engineers in the world have yet to design a closed life support system. Designing one appears to be a fiendishly difficult goal to achieve.

One reason for this difficulty is designing failure tolerances. What are these? A *failure tolerance* means that a failure in one subsystem will not result in a failure of the whole system or a situation that threatens the lives of the crew. That is why an LSS must be fail-safe or fail-operational. *Fail-safe* means the system has the ability to sustain a failure but still retain the capability of keeping the crew and mission safe, whereas *fail-operational* means the system can fail, but even after it has failed, it can still meet the capability of retaining full operational capability [4, 5, 6].

To ensure these requirements are met, LSS engineers must consider myriad factors, including power consumption, habitable volume, storage requirements, mass ratio, and maintenance requirements. Each of these factors has an impact (☐ Table 3.2) on the design of an LSS, and to determine the impact of each factor, it is necessary to conduct what are called trade studies. These studies use simple models to identify the effect of one factor or several factors against specific LSS technologies. For example, a subsystem might perform at a very high level of efficiency, but if that system is the size of a truck, then it obviously can't be used on a spacecraft where space is limited. Similarly, there might be a super-efficient subsystem that does a great job, but if it needs maintenance every 3 days, then it will require some rethinking.

☐ **Table 3.2**   Mission factors and life support system impacts

Mission factor	Impact on design of life support system
Number of crew	More astronauts means more consumables are required
Mission length	A long mission will require more consumables than a short mission. A long mission will also require higher reliability and longer times between maintenance and repair
Spacecraft leakage	The more a spacecraft leaks, the greater the demand on the atmosphere control system
Resupply capability	The longer the mission, the greater the demand for reliability and life support system closure
Power consumption	There will always be limited power, and this has to be managed to provide power to the myriad subsystems and assemblies
Volume	Spacecraft have limited volume, so systems and subsystems have to be designed to be as small as possible
Gravity	Systems and subsystems must be designed to work in various gravity effects – microgravity, one-sixth gravity (lunar), and one-third gravity (Mars)
Contamination	Spacecraft systems produce contamination, and the LSS must be designed to deal with these contaminants
In situ resource utilization (ISRU)	This will reduce the demand on the LSS at the destination

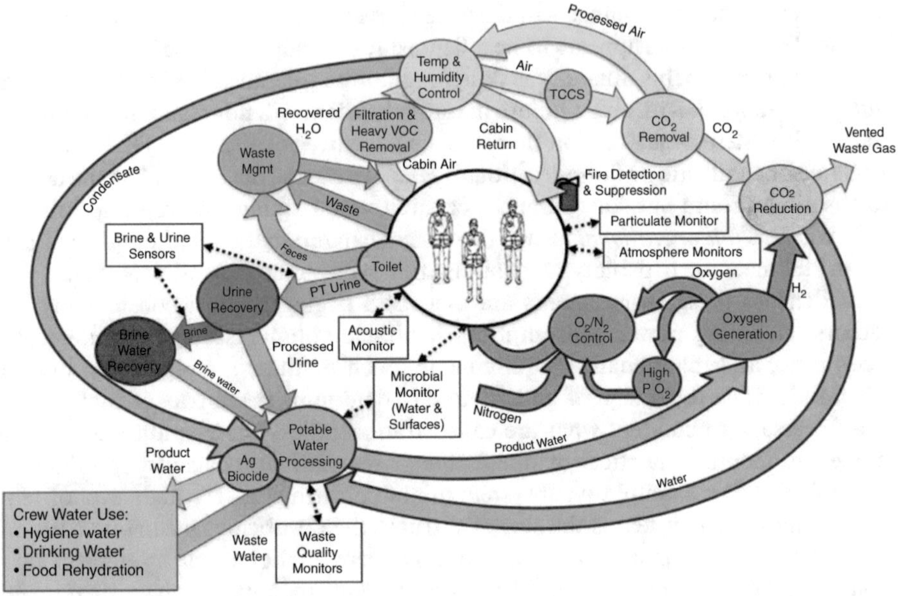

■ **Fig. 3.5** The life support system of the International Space Station. Credit: NASA

To give you an idea of the difficulty faced by engineers when considering all these factors, take a look at ■ Fig. 3.5, which depicts the interaction of all the subsystems of the LSS on board the International Space Station (ISS). A more detailed analysis of these factors is presented in ■ Table 3.3, which considers all the interfaces, each of which must be considered in these trade studies (■ Fig. 3.6).

Crew Requirements

Obviously, the primary goal of the LSS, regardless of what category, is sustaining life. To sustain life, an LSS must meet human physiological requirements, because they are the drivers in the design process. Unfortunately for LSS engineers, humans can only survive under very precise conditions [7, 8]. Consider the following facts: most humans will expire if deprived of oxygen for 4 minutes, most will die if deprived of water for 3 days, and most cannot survive after 30 days without food. Not only that, but humans require a comfortable habitat, and they produce waste that also has to be taken care of. So let's take a look at some of these crew requirements. We'll begin with the basics of food and water and waste. These consumables are depicted in ■ Fig. 3.7.

Environmental Requirements: Radiation

In the context of spaceflight, the environment must provide a comfortable atmosphere, artificial gravity, and protection from radiation. Achieving the first of these requirements has already been accomplished, but achieving the second requirement

◘ Table 3.3 Life support interfaces (BVAD)[1]

Life support interfaces	Description	Life support system interfaces
Crew	The crew interface interacts with all life support sub-systems and interfaces. It accounts for all metabolic inputs and outputs from crewmembers. Historically and likely in the near term (until other animals or plants are included in the mission in large scales), crewmembers are the foremost consumers of life support commodities and the primary producers of waste products	All
Environmental monitoring and control	The environmental monitoring and control (EMC) interface provides information on the chemical and biological status of the crew habitat. This includes trace and major constituent composition of air and water, smoke detection, and microbial content of air, water, and surfaces. The information is used to control proper functioning of the life support system, as well as indicate off-nominal events	All
Extravehicular activity support	The extravehicular activity (EVA) support interface provides life support consumables including oxygen, water, and food for all suited activities, as well as carbon dioxide and waste removal. Suits may be employed for launch, entry, and abort (in case of cabin depressurization), nominal or contingency EVA in a weightless environment, emergency return from a human mission beyond low Earth orbit, and surface EVA operations on the Moon and Mars	Air, habitation, waste, water, EMC, crew, food, power, thermal
Food	The food interface provides the crew with prepackaged food products or commodities requiring some level of preparation or processing and includes the stowage systems necessary for these items. If an advanced life support system were to include a biomass subsystem, the food system would also receive harvested agricultural products and process them into an edible form	Air, habitation, waste, water, EMC, crew, EVA support, food, power
Habitation	The habitation interface is responsible for crew accommodations and human engineering. The packaging and preparation and storage of crew supplies include the galley layout and food supplies, clothing management systems, fire suppressant, gas masks, hygiene stations and supplies, housekeeping and related supplies, and other functions related to configurable crew living. This technology area is responsible for implementing the hardware resulting from human factor requirements	Air, waste, water, EMC, crew, EVA support, food, power, radiation protection, radiation

(continued)

3

◘ Table 3.3 (continued)

Life support interfaces	Description	Life support system interfaces
In situ resource utilization	The in situ resource utilization interface provides life support commodities such as gases, water, and regolith from local planetary materials for use throughout the life support system	Air, water, EMC, crew, power, radiation protection
Medical systems	Under nominal conditions, medical systems would generally have an inconsequential impact on the life support systems, but if an event should occur that causes illness or injury, the impacts on the life support system could be drastic. This includes medical and metabolic monitoring of the crew during EVAs. Gases may be required for hyperbaric treatment and respiratory therapy or to provide oxygen for certain medical procedures while controlling flammability risks in the cabin. Additional water may be required, and waste could be generated that might not be allowed to be stored, processed, or recycled like waste from nominal activities	Air, waste, water
Power	The power interface provides the necessary energy to support all equipment and functions within the life support system. It may also provide resources like fuel cell product water to the life support system	All
Propulsion	The propulsion interface may provide resources such as oxygen and cooling evaporant to the life support system and thermal control system	Air, waste, water, EMC, EVA support, thermal
Radiation protection	The radiation protection interface includes systems design to provide the crew protection from environmental radiation. The life support system could provide some useful contribution to radiation protection, especially in the form of water or waste products. The radiation protection interface also provides sensors and other predictive measures for solar particle events, so the crew might seek shelter from such an event	Habitation, waste, water, crew, food, ISRU, power
Thermal	The thermal interface is responsible for maintaining cabin temperature and humidity (unless controlled jointly with other atmosphere revitalization processes) within appropriate bounds and for collection and removal of the collected waste heat from crew, equipment, and the pressurized volume to the external environment. Note: equipment to remove thermal loads from the cabin atmosphere normally provides sufficient air circulation. Thermal interface work is conducted under the Thermal Control System Development for Exploration Project	Air, habitation, waste, water, EMC, crew, EVA support, food, power

[1]Adapted from NASA's Baseline Values and Assumptions Document. NASA/TP-2015–218570. March 2015

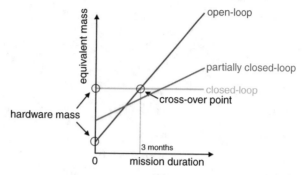

◘ **Fig. 3.6** Example of the type of trade studies that must be completed when trying to design a life support system. Credit: NASA

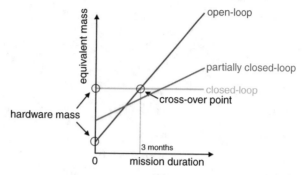

Life Requirements on Earth and in Space

Item	On Earth		In Space	
	kg per person per day[1]	gallons per person per day	kg per person per day[2]	gallons per person per day
Oxygen	0.84		0.84	
Drinking Water	10	2.64	1.62	0.43
Dried Food	1.77		1.77	
Water for Foood	4	1.06	0.80	0.21

◘ **Fig. 3.7** Graphical representation of the significant mass inputs and outputs. Credit: NASA [8]

may take quite a while. As for the third requirement, this may be something of a mission killer when we start talking about missions to Mars. We discuss artificial gravity in ▶ Chap. 9 and radiation protection is discussed in ▶ Chap. 7. An annual dose of radiation here on Earth is 2 mSv, whereas an astronaut spending 6 months on the ISS will be exposed to 80 mSv. But an astronaut taking a trip to Mars? Well, any such astronaut will exceed their career radiation dose limit. As a comparative reference, the dose limits for a 1-year mission are provided in ◘ Table 3.4.

Environmental Requirements: Metabolic Rates

Metabolic activity is the result of converting food to energy by the astronauts. This process exerts an effect not only on air revitalization and heat production but also on water utilization, waste production, and power consumption. When calculating metabolic rates, LSS engineers rely on equations used by NASA in their Human Integration Design Handbook (HIDH) document [9]. In this equation, which is

3

◻ Table 3.4 Sample career effective dose limits for 1-year missions for a 3% REID and estimates of average life loss if death occurs

Age at exposure	E (mSv) for a 3% REID (average life loss per death)	
	Male	**Female**
30	620 (15.7)	470 (15.7)
35	720 (15.4)	550 (15.3)
40	800 (15.0)	620 (14.7)
45	950 (14.2)	750 (14.0)
50	1150 (12.5)	920 (13.2)
55	1470 (11.5)	1120 (12.2)

provided below, crew time is expressed as CM-h and crew days as CM-d, and crew-member mass is a range from the 95th percentile American male to 5th percentile Japanese female.

Human Metabolic Rate Equation *males* > 19 years old

$$\left(\frac{622 - 9.53 \times \text{age}(\text{yrs}) + 1.25(15.9 \times \text{mass}(\text{kg}) + 539.6 \times \text{ht}(\text{m}))}{0.238853 \times 10^3} = \text{Energy} \ \frac{\text{MJ}}{\text{CM} - \text{d}} \right)$$

Human Metabolic Rate Equation *females* > 19 years old

$$\left(\frac{354 - 6.91 \times \text{age}(\text{yrs}) + 1.25(9.36 \times \text{mass}(\text{kg}) + 726 \, \text{ht}(\text{m}))}{0.238853 \times 10^3} = \text{Energy} \ \frac{\text{MJ}}{\text{CM} - \text{d}} \right)$$

By applying this equation, it is possible to calculate metabolic rate for various activities (◻ Table 3.5). Armed with this data, LSS engineers can then predict the demands on the life support system.

As you can see in ◻ Table 3.5, there is a lot of metabolic data (taken from NASA's HIDH document) that must be considered. This table gives you a snapshot of the metabolic costs of a select number of activities. The next step in calculating energy cost is determining the time taken by crewmembers to complete routine daily activities, as depicted in ◻ Table 3.6.

◻ **Table 3.5** Metabolic rates for various activities

1 Crewmember activity description	2 Duration of activity (hr)	3 Dry heat output kJ/hr	4 Wet heat output kJ/hr	5 Total heat output rate kJ/hr	6 Water vapor output kg/min	7 Sweat runoff rate kg/min	8 Oxygen consumption kg/min	9 CO2 output kg/min
Sleep	8	224	92	317	6.3	0.00	3.6	4.55
Nominal	14.5	329	171	500	11.77	0.00	5.68	7.2
Exercise 0–15 min at 75% VO$_2$ max	0.25	514	692	1206	46.16	1.56	39.4	49.85
Exercise 15–30 min at 75% VO$_2$ max	0.25	624	2351	2974	128.42	33.52	39.4	49.85
Recovery 0–15 min post 75% VO$_2$ max	0.25	568	1437	2005	83.83	15.16	5.68	7.2
Recovery 15–30 min post 75% VO$_2$ max	0.25	488	589	1078	40.29	0.36	5.68	7.2
Recovery 30–45 min post 75% VO$_2$ max	0.25	466	399	865	27.44	0.00	5.68	7.2
Recovery 45–60 min post 75% VO$_2$ max	0.25	455	296	751	20.4	0.00	5.68	7.2
Total/day	24	7351	4649	12000	1.85	0.08	0.82	1.04

Environmental Requirements: Nutrition

To maintain optimum physiological function, astronauts must be supplied with a regular supply of energy and nutrients. The primary nutrients are carbohydrates, fats, and protein. The ISS menu provides approximately 50% of calories from carbohydrates, 20% of calories from protein, and 30% of calories from fat [10]. But providing calories and nutrients is just one of myriad considerations when it comes to deciding how best to support nutritional requirements. History has shown that

3

Table 3.6   Crewmember daily routine activities

Activity	Weekday (Cm-h /CM-d)	Weekend Day (CM-h / CM-d)
Daily planning conferences	0.5	0.0
Daily plan review/report preparation	1.0	0.0
Work preparation	0.5	0.0
Scheduled system utilization operations	6.5	0.3
Meals	3.0	3.0
Housekeeping	0.0	2.0
Post-sleep	0.5	0.5
Exercise, hygiene	2.5	2.5
Recreation	0.0	6.0
Presleep	1.0	1.0
Sleep	8.5	8.5
Total	24	24

NASA/TP-2015–218570. Life Support Baseline Values and Assumptions Document

astronauts rarely consume their required daily ration of calories. This is a problem because microgravity reduces muscle mass, and protein synthesis is required to maintain muscle mass, but inadequate energy intake is related to reduced protein synthesis [11, 12].

Then there is the question of bone mass. Several studies have shown that adequate energy, protein, and vitamin D intake are required to maintain bone health during several months on orbit. But the ISS diet is generally higher than it should be in sodium (5300 mg/d), and research has shown that high sodium has bone-resorbing effects during exposure to microgravity [13, 14]. Vision is another factor to consider when deciding on an optimal nutritional intake for astronauts. That is because the vision-related issue astronauts suffer from time to time on orbit may be related to serum folate levels, vitamin B-12 intake, and homocysteine [15].

Physicochemical Life Support Systems

A space-based physicochemical LSS is one in which the astronaut is the only biological component. Such a system, such as the one on the ISS (see ► Chap. 5), can meet the tasks of managing the atmosphere, managing water, and managing waste (▪ Fig. 3.8). We'll take a look at each of these requirements here.

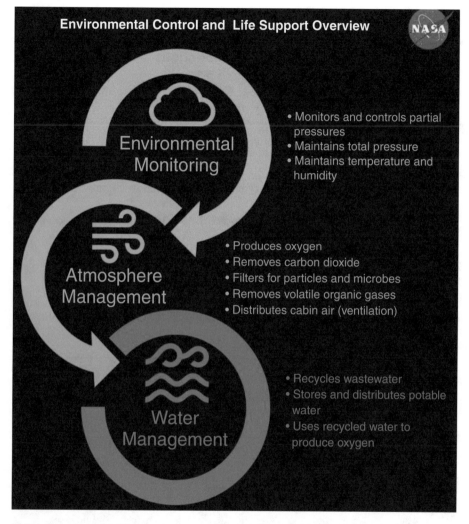

□ Fig. 3.8 The ISS LSS is a physicochemical LSS: the primary functions are depicted in this diagram. Credit: NASA

Atmosphere Management

This function can be divided into the separate functions of controlling and supplying the atmosphere, maintaining temperature and humidity, monitoring the atmosphere, ventilating the atmosphere, and fire detection and suppression.

Atmosphere Control and Supply Functions
- Maintain atmospheric pressure.
- Maintain partial pressures of oxygen, nitrogen, and carbon dioxide.
- Regulate atmospheric pressure and partial pressures.
- Store oxygen and nitrogen.

3

- Distribute oxygen and nitrogen.
- Operate autonomously with limited crew intervention.

Temperature and Humidity Control Functions

- Maintain temperature between 18°C and 26°C.
- Maintain humidity between 25% and 70%.
- Monitor trace contaminants and particulates.
- Control microbial levels.
- Operate autonomously with limited crew interaction.

Atmosphere Monitoring

- Monitoring, identification, and quantifying of volatile organic compounds (VOC).
- Audible and visual alarms to alert crewmembers when concentrations of VOC's exceed maximum allowable levels.
- Microbial decontamination capabilities.
- Control of contamination events.

Cabin Ventilation

- Ventilation to ensure thermal gradients maintained and contaminant buildup reduced
- Ventilation to ensure cooling

Air Revitalization

- Maintaining carbon dioxide concentration
- Reducing carbon dioxide concentration
- Generating oxygen

Air Revitalization Technologies

Function	Technology
Carbon dioxide concentration	Lithium hydroxide
	Four-bed molecular sieve
	Solid amine water desorption
Carbon dioxide reduction	Activated charcoal
	Sabatier
	Bosch
Oxygen generation	Water vapor electrolysis
	Solid polymer water electrolysis

An explanation of all the candidate technologies for achieving all the functions of atmosphere management is beyond the scope of this chapter (see ▶ Chap. 5 for a more detailed discussion of the ISS LSS subsystems and assemblies), so we will focus on select technologies such as the four-bed molecular sieve (4BMS) technology used on ISS (see ◘ Fig. 3.9).

The 4BMS, which is a mature technology, having first been used on Skylab, is a regenerable means of removing carbon dioxide and of maintaining carbon dioxide concentration at toler-

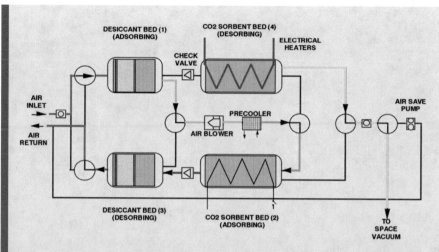

▫ **Fig. 3.9** Schematic of the four-bed molecular sieve that is part of the carbon dioxide removal assembly (CDRA) on the ISS [16, 17]. This diagram shows one half-cycle of operation as follows: (1) Humid cabin air flows through the adsorbing desiccant bed. (2) Air is then directed through a blower and precooler. (3) The air, which is now dry, is directed through a zeolite sorbent bed which is where carbon dioxide is adsorbed. (4) The air is then directed through the desorbing desiccant bed

able levels. A regenerable system such as the 4BMS is different from a non-regenerable process such as lithium hydroxide, which you would find in an open-loop LSS (remember, the ISS LSS is classed as a partially closed-loop/physicochemical LSS). The use of lithium hydroxide, the use of which goes back to the Mercury era, removes carbon dioxide from the spacecraft atmosphere by directing carbon dioxide-rich air through a canister packed with lithium hydroxide granules. It requires about 2 kilograms of lithium hydroxide to remove 1 kilogram of carbon dioxide (the daily output by a crewmember), so this system works well for short-duration missions but not so well for missions longer than 2 weeks.

Obviously, the use of lithium hydroxide won't help LSS engineers close the loop in a physicochemical LSS, but there is a system that will: the Sabatier process. Developed by Nobel Prize-winning French chemist Paul Sabatier in the early 1900s, the Sabatier process uses a catalyst to react carbon dioxide and hydrogen to produce water and methane, thereby closing the oxygen and water loops. The system, which is currently in operation on the ISS, is integrated with the oxygen generating system (OGS).

Bioregenerative Life Support Systems

While this book is being written, in mid-2020, space agencies are in the "picnic approach" in LSS evolution, with most supplies needed for the mission being delivered by cargo resupply missions. This is because the ISS relies on the physicochemical system discussed in the previous section. But if astronauts are finally to embark on missions beyond Earth orbit, regenerative life support systems must be devel-

3

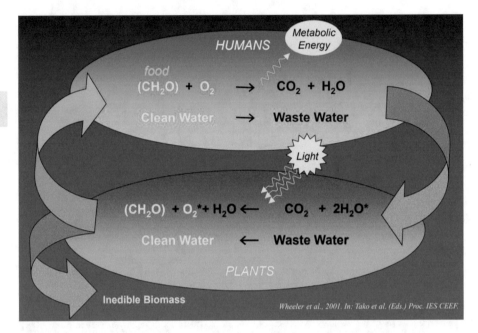

◘ Fig. 3.10   For life support systems to become truly closed, a bioregenerative system is the way to go. Credit: NASA/Wheeler et al.

oped. The key difference between a physicochemical LSS and a bioregenerative LSS (BLSS) is that a BLSS can produce food (◘ Fig. 3.10) in addition to all the functions that can be accomplished by a physicochemical system.

In NASA's Advanced Life Support Program, BLSS research is being conducted to develop technologies that can help optimize biomass productivity (◘ Fig. 3.11), recycle liquid and solid wastes, and build systems that will enable long-term operations [18, 19, 20]. Once these technologies have been developed, the transition from the picnic approach to permanent recycling and independence from the resupply chain will be complete.

As we shall see in ▶ Chap. 8, growing food is the biggest technology gap in developing a true BLSS, but experiments conducted on the ISS seem to be on the right track, and researchers are slowly but surely gaining an understanding of some of the basic processes of how plants grow in space. For example, such experiments have revealed the underlying mechanisms of *circumnutation*, a circular movement of a growing stem. In space, studies of *Arabidopsis* have revealed that patterns of root waving during sprouting are similar to patterns observed on Earth, thereby showing that gravity does not exert a major effect on patterns of root growth.

But the function of growing plants is not just to provide food for the crew. Oxygenic photoautotrophic organisms also happen to be of benefit to the management of the atmosphere, the management of water, and the management of waste. These plants not only produce food but also use up carbon dioxide created by the astronauts, which is a win–win. Not only that, but plant transpiration can be recovered by condensation and then directed into the water recovery system for

Fig. 3.11 The VEGGIE experiment, being tended to by Scott Tingle. Among the plants that have been grown in this experiment are red lettuce, red Russian kale, Wasabi mustard, and extra dwarf pak choi. Credit: NASA [21, 22]

processing before being used as potable water again. Plant biological processes can also give a helping hand to the waste processing system by decreasing the mass and volume of biodegradable elements.

So why don't we have such a system on board the ISS? Engineering such a system is deviously complex, as we shall see when we take a look at MeLiSSA in ▶ Chap. 8. Building an efficient BLSS requires the very, *very* careful selection of plants that can perform all those aforementioned LSS functions, and these plants must be ecologically compatible not only with all the other organisms in the system but also with the crew. Compounding the challenges in achieving this compatibility is the absence of natural forces such as gravity, which means that all systems must be under scrupulous control mechanisms. In 2020, a fully developed BLSS is a long way away, but there is a development path toward achieving such a system, known as the controlled ecological life support system (CELSS). CELSS is a pathfinder program working toward the Holy Grail of a closed LSS.

Controlled Ecological Life Support System

The development and validation of CELSS technologies are still ongoing. One aspect of that development is including humans in the loop. Such studies (most of which have focused more on group interaction – or lack of it! – than life support system function) have been carried out on a fairly regular basis over the years with mixed results:

3

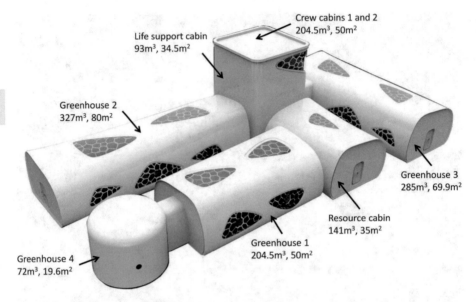

Crew cabins 1 and 2
204.5m³, 50m²

Life support cabin
93m³, 34.5m²

Greenhouse 2
327m³, 80m²

Greenhouse 3
285m³, 69.9m²

Resource cabin
141m³, 35m²

Greenhouse 1
204.5m³, 50m²

Greenhouse 4
72m³, 19.6m²

◘ **Fig. 3.12**   The modules in which four crewmembers spent 180 days testing a CELSS. The habitat was located at the Space Institute of Southern China, Shenzhen. Oxygen partial pressure ranged from 18.6 to 26.7 kPa, and carbon dioxide concentration ranged from 300 to 700 ppm. The test took place between 17 June 2016 and 14 December 2016. Credit: Chinese National Space Administration [23, 24, 25, 26]

- The HUBES (Human Behavior in Extended Spaceflight) study, September 1994 to January 1995 in Bergen, Norway – generally very successful
- The SFINCSS-99 (Simulated Flight of International Crew on Space Station) – blighted by sexual harassment, poor team cohesion, arguments with mission control, etc.
- The Mars500 boondoggle – a remarkably unremarkable study that revealed nothing that couldn't have been gleaned from a read of any one of Shackleton's logbooks!

A more recent study, which focused more on life support and less on crew discord, conflict, and strife, was a 180-day Chinese CELSS study [23, 24, 25, 26] conducted using a closed-loop system during a mid-mission simulation of a Mars mission. The aim was to study the physiological and psychological effects when relying on a closed-loop system, but we'll focus on the physiological effects here. The habitat (◘ Fig. 3.12) comprised six interconnected modules in which life support systems were controlled automatically.

The habitat provided the Chinese crew of four (3 males and 1 female aged 34.2 ± 6.6 years, weight 64.5 ± 6.1 kilograms) with a luxurious 1340 m³ of habitable volume (the habitable volume of the ISS by comparison is 931 m³), of which a whopping 888.5 m³ comprised greenhouses. Why so much space dedicated to greenhouses? Because this crew had to meet their food needs autonomously. During their time inside the habitat, the crew cultivated 25 types of plants, includ-

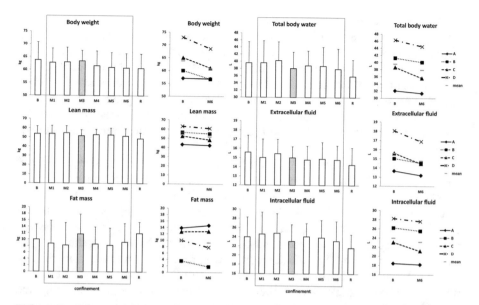

⬛ Fig. 3.13 Select results of the Chinese 18-day CELSS mission. Credit: Chinese National Space Administration [23, 24, 25, 26]

ing potatoes, soybeans, lettuce, cabbage, cherry radish, and strawberries. They kept themselves busy maintaining the LSS, looking after plants and general housekeeping duties. During the simulated missions, 100 percent oxygen and 99 percent water were regenerated in addition to 70 percent of food being grown (the remaining 30% was stored prior to the start of the mission). The daily energy intake (35% fat, 50% carbohydrate, 15% protein) of the crew was ~2600 kcal at the beginning of the mission and 2000 kcal at the end.

So what happened to the crew? Body weight decreased slightly (64 ± 7 kilograms at the beginning vs. 61 ± 6 kilograms at the end), lean mass decreased (54 ± 8 kilograms at the beginning vs. 52 ± 8 kilograms at the end), and vitamin D levels fell (⬛ Fig. 3.13). Blood counts, renal function, and metabolism were all within nominal states following the end of the mission.

So, the Chinese study represents a good start in the development of a closed LSS, but does BLSS technology have to progress before we have such a system integrated on a spaceship? There are a number of challenges ahead, some of which are outlined here.

One such challenge is to create detailed environmental control algorithms to improve system stability for tracking contaminants created by pathogenic microorganisms [18]. A second hurdle is defining exactly how much space and resources are required per crewmember for a BLSS/CELSS [25]. A third issue is to develop the capability to track plant diseases and how to control these diseases. This is important because a couple of crop-wrecking diseases can quickly result in loss of mission and loss of crew! Then there is the question of how to measure the balance of nutrient cycling between food production and waste mineralization. For example, what are the limits for waste composition for each plant?

Another area that has not been subject to much research is the use of aquatic plants as food. Some aquatic plants are particularly fast growing and produce high yields. Then there is the challenge of automation. Tending to plants is rather time-intensive, so integrated control systems must be developed that are capable of not only maintaining nominal conditions but also responding automatically to unexpected events. Let's add to this list of challenges by adding the requirement to determine nutrient management for crops and defining the processes required for converting raw food into edible food (these processes will no doubt be affected by microgravity) [27].

For many of these challenges, modeling can be used as a first step to test and analyze system dynamics in response to the microgravity environment, but ultimately because of the complexity of the natural feedback mechanisms in biological systems, it is impossible to mathematically describe how a complex ecosystem such as a CELSS would function in reality. Think about all the permutations involved – mass transport such as advection and diffusion, mass balance, solar energy input, temperature and humidity variations, chemical processes such as reaction kinetics, enzymatic reactions, hydrolysis, ion exchange, and the list goes on. And on. And we're just talking about factors under nominal conditions. Once you've figured how all these factors function nominally, you then have to predict what will happen in myriad emergency scenarios. So, modeling can only do so much, no matter how structurally dynamic those models are.

Ultimately, the only way to reliably test is to build a prototype and test this in space. And remember, these tests will fall under the category of long-term experiments, because a typical manned Mars mission is in the order of 30 months or more. This time span is also necessary to capture true failure rates over the system lifespan. But conducting these sorts of experiments with human operators is time-intensive, and astronauts on board the ISS just don't have the time to do this because they're too busy tending to the myriad maintenance issues of the LSS.

Given that the lifespan of the ISS extends to 2028, how will LSS engineers be able to test the technologies required to develop a full-scale CELSS/BLSS? To begin with, they will have to turn to terrestrial-based experiments. Unfortunately, there are only seven facilities on the planet that can support integrated full-scale LSS testing. And these facilities are big and extremely expensive and require a lot of time and work to operate. How about testing small-scale CELSSs? Well, this option has been pursued by ecologists for some time now, and these compact, controllable, replicable ecosystems, or *microcosms* as they are called, have proven to be a good way of testing complex system behavior on a budget. The problem with using these microcosms is that they cannot serve as analogs for the real thing, as you cannot trust that what happens in a scaled model is representative of what will happen in a full-scale system. That is because processes change as scale changes! And because processes change as scale changes, it is not possible to reliably apply data from a microcosm to a large-scale CELSS.

In short, quantitative and systematic extrapolation will just not work. Instead, the goal of any terrestrial experiment must be to ensure functional similarity in ecological relationships within the CELSS. What does this mean? Experiments must be long enough to ensure all processes can be observed. It means that all factors – water exchange, nutrient concentrations, water depth, light attenuation, and ecological complexity, to name just a few – are controlled. To begin with, many of these parameters can be tested in smaller closed ecological systems that can serve as analogs. The purpose of these systems would be to generate baseline data that can ultimately be applied to the full-scale systems that must eventually be tested for the reasons stated previously.

The science of ecology is progressing in leaps and bounds, thanks to the observations made over decades of research at integrated test facilities. But ultimately, the complexity of ecological system dynamics means that even with the most advanced algorithms and modeling, it is not yet possible to predict how systems modeled on the ground will work in space. To further develop the CELSS to the point where we eventually have a working BLSS, what needs to be done? Well, engineers have to develop systems that work with a higher reliability, the closed ecosystem must have greater control, biological stability must be improved, and all these factors must be tested at full-scale integrated LSS test facilities.

Key Terms
- Baseline Values and Assumptions Document (BVAD)
- Bed Molecular Sieve (BMS)
- Bioregenerative Life Support System (BLSS)
- Carbon Dioxide Removal Assembly (CDRA)
- Closed Ecological Life Support System (CELSS)
- Environmental Monitoring and Control (EMC)
- Extravehicular Activity (EVA)
- International Space Station (ISS)
- Oxygen Generating System (OGS)
- Volatile Organic Compounds (VOC)

❓ Review Questions

1. What is the difference between an open-loop and a closed-loop life support system?
2. Which type of life support system is on the ISS?
3. What is the difference between fail-safe and fail-operational?
4. List four interfaces of the EVA system.
5. What is EMC?
6. List four functions of the Atmosphere Control and Supply System.
7. What is CDRA?
8. How is 4BMS used?
9. What is the Sabatier process?
10. Describe four challenges in designing a CELSS.

3

References

1. ESM GD. *Advanced life support equivalent system mass guidelines document.*" NASA TM-2003-212278.
2. NASA. "Exploration Life Support Requirements Document." JSC-65527A, National Aeronautics and Space Administration, Lyndon B. Johnson Space Center, Houston, Texas. [ELS RD (June 2008)]
3. Anderson, M. S., Ewert, M. K., & Keener, J. F. (2018, Jan). *Life support baseline values and assumptions document,* NASA/TP2015–218570/REV1.
4. NASA. (2007). "NASA Space Flight Human-System Standard Volume 1, Revision A: Crew Health." NASA-STD-3001, Volume 1, Revision A, approved 7-30-2014. NASA. HIDH - "Human integration design handbook." NASA/SP-2010-3407/REV1, approved 6-5-2014. NASA-STD-3001.
5. Escobar, C. M., Nabity, J. A., & Klaus, D. M. (2017). *Defining ECLSS robustness for deep space exploration,* 47th international conference on environmental systems, (submitted for publication).
6. Levri, J. A., Drysdale, A. E., Ewert, M. K., Hanford, A. J., Hogan, J. A., Joshi, J. A., & Vaccari, D. A. (2003). *Advanced life support equivalent system mass guidelines document,.* NASA/TM-2003-212278.
7. Lane, H. W., Sauer, R. L., & Feeback, D. L. (2002). *Isolation: NASA experiments in closed-environment living, advanced human life support enclosed system final report* (Vol. 104 Science and Technology Series). American Astronautical Society, copyright.
8. NASA. (2010). *Exploration life support baseline values and assumptions document.* Houston, TX: JSC-64367 Rev B. National Aeronautics and Space Administration, Lyndon B. Johnson Space Center.
9. NASA, *Human Integration Design Handbook (HIDH),* NASA/SP-2010-3407/REV1, 06-05-2014.
10. Lane, H. W., Bourland, C. T., Pierson, D., Grigorov, E., Agureev, A., & Dobrovolsky, V. (1996). *Nutritional requirements for international space station missions up to 360 days.* Houston: JSC-28038, National Aeronautics and Space Administration, Lyndon B. Johnson Space Center.
11. Lane, H. W., & Schoeller, D. (2000). *Nutrition in spaceflight and weightlessness models.* Boca Raton: CRC Press.
12. Stein, T. P., Leskiw, M. J., Schluter, M. D., Donaldson, M. R., & Larina, I. (1999). Protein kinetics during and after long-duration spaceflight on MIR. *The American Journal of Physiology, 276,* E1014–E1021.
13. Buehlmeier, J., Frings-Meuthen, P., Remer, T., Maser-Gluth, C., Stehle, P., Biolo, G., & Heer, M. (2012). Alkaline salts to counteract bone resorption and protein wasting induced by high salt intake: results of a randomized controlled trial. *The Journal of Clinical Endocrinology and Metabolism, 97,* 4789–4797.
14. Zwart, S. R., Hargens, A. R., & Smith, S. M. (2004). The ratio of animal protein intake to potassium intake is a predictor of bone resorption in space flight analogues and in ambulatory subjects. *The American Journal of Clinical Nutrition, 80,* 1058–1065.
15. Zwart, S. R., Gibson, C. R., Mader, T. H., Ericson, K., Ploutz-Snyder, R., Heer, M., & Smith, S. M. (2012). Vision changes after spaceflight are related to alterations in folate- and vitamin B-12-dependent one-carbon metabolism. *The Journal of Nutrition, 142,* 427–431.
16. Coker, R. F., & Knox, J. C. *Predictive modeling of the CDRA 4BMS,* 46th International Conference on Environmental Systems, Vienna, Austria, ICES-2016-92, 2016.
17. Knox, J. C., & Stanley, C. M., *Optimization of the Carbon Dioxide Removal Assembly (CDRA-4EU) in support of the international space system and advanced exploration systems,* 45th international conference on environmental systems, Bellevue, WA, ICES-2015-165, 2015.
18. Barta, D. J., Castillo, J. M., & Fortson, R. E. *The biomass production system for the bioregenerative planetary life support systems test complex: preliminary designs and considerations.* SAE paper 1999-01-2188, 29th international conference on environmental systems, Society of Automotive Engineers, Warrendale, PA, 1999.

19. Wheeler, R. M., Mackowiak, C. L., Stutte, G. W., Yorio, N. C., Ruffe, L. M., Sager, J. C., & Knott, W. M. (2006). *Crop production data for bioregenerative life support: Observations from NASA's Kennedy Space Center*. Paris: COSPAR Abstract F4.1-0010-06, Committee on Space Research, International Council for Science.

20. Wheeler, R. M., Sager, J. C., Prince, R. P., Knott, W. M., Mackowiak, C. L., Stutte, G. W., Yorio, N. C., Ruffe, L. M., Peterson, B. V., Goins, G. D., Hinkle, C. R., & Berry, W. L. (2003). *Crop production for advanced life support systems – observations from the Kennedy Space Center Breadboard Project*. NASA-TM-2003-211184. National Aeronautics and Space Administration, John F. Kennedy Space Center, FL.

21. Godia, F., Albiol, J., Montesinos, J. L., Perez, J., Vernerey, A., Pons, P., & Lasseur, C. (1997). *MELISSA pilot plant: A facility for the demonstration of a biological concept of a life support system* (pp. 873–878). European Space Agency Publications, ESA SP 400.

22. Lasseur, C., Brunet, J., De Weever, H., Dixon, M., Dussap, G., Godia, F., Leys, N., Mergeay, M., & Van Der Straeten, D. (2010). MELiSSA: the European project of closed life support system. *Gravitational and Space Research, 23*(2), 3.

23. Yuan, M., Custaud, M.-A., Xu, Z., Wang, J., Yuan, M., Tafforin, C., Treffel, L., Arbeille, P., Nicolas, M., Gharib, C., Gauquelin-Koch, G., Arnaud, L., Lloret, J.-C., Li, Y., & Navasiolava, N. (2019). Multi-system adaptation to confinement during the 180-day Controlled Ecological Life Support System (CELSS) experiment. *Frontiers in Physiology, 10*, 575.

24. Li, T., Zhang, L., Ai, W., Dong, W., & Yu, Q. (2018). A modified MBR system with post advanced purification for domestic water supply system in 180-day CELSS: construction, pollutant removal and water allocation. *Journal of Environmental Management, 222*, 37–43.

25. Xie, B., Dong, C., & Wang, M. (2016). How to establish a bioregenerative life support system for long-term crewed missions to the moon or mars. *Astrobiology, 16*(12), 925–936.

26. Liu, H. (2014). Bioregenerative life support experiment for 90-days in a closed integrative experimental facility LUNAR PALACE 1, 40th COSPAR Scientific Assembly, Vol. 40.

27. Saltykov, M., Bartsev, S. I., & Lankin, P. (2012). Stability of Closed Ecology Life Support Systems (CELSS) models as dependent upon the properties of metabolism of the described species. *Advances in Space Research, 49*(2), 223–229.

Suggested Reading

NASA. (2018, Jan). Life support baseline values and assumptions document, NASA/TP2015-218570/REV1.

NASA. Human Integration Design Handbook (HIDH). NASA/SP-2010-3407/REV1, 06-05-2014.

Evolution and Development of Life Support Systems

STS-116 Mission specialist Christer Fugle-
sang is helped with his Advanced Crew Escape
Suit before entering Space Shuttle Discovery.
Credit: NASA/Amanda Diller

© Springer Nature Switzerland AG 2020
E. Seedhouse, *Life Support Systems for Humans in Space*,
https://doi.org/10.1007/978-3-030-52859-1_4

Contents

Learning Objectives

After reading this chapter, you should be able to:

- Describe the evolution of life support systems and subsystems across the Mercury, Gemini, Apollo, and Space Shuttle Programs
- Explain the function of each spacecraft life support subsystem in the Mercury, Gemini, Apollo, and Space Shuttle Programs
- Describe the life support system issues experienced during specific missions during the Apollo Program
- Describe the function of the components of the Extravehicular Mobility Unit and the operational characteristics of the pressure garment assembly
- Explain how the portable life support system performed its tasks
- List the key portable life support system specifications
- Describe the physiological measurements monitored during the Mercury, Gemini, and Apollo missions
- List some of the life support issues encountered during the Apollo program, and list the corrective actions of each

Introduction

The first life support system designed to support a cosmonaut was that flown on Yuri Gagarin's Vostok spacecraft. But this wasn't the first life support system to be flown in space. For years preceding Gagarin's flight, animals had been sent into space. Before Gagarin's flight, there was Laika, a mongrel pup, who flew into orbit on board Sputnik 2. Because of the claustrophobic confines of the Sputnik 2 capsule, the spacecraft's crewmember could not be heavier than 7 kilograms. Ten dogs that met the weight requirements were selected, but Laika was judged the calmest and most photogenic!

As part of her training, Laika was taught to remain still in increasingly confined cages in the weeks leading up to the mission. Unfortunately, due to the Soviet government's desire for a quick success, Laika's life support system was not as rigorously tested as it should have been. She made it to space alive, but she didn't have much time to enjoy her time in space. The thermal control system failed, resulting in Laika expiring from overheating after just 5 hours on orbit.

Before Laika, there were Tsygan and Dezikin, who flew suborbital flights. Other countries had also launched animals into space, including the French, who launched a cat by the name of Felicette, and the Americans, who launched a number of chimpanzees, the most famous being Ham. These animals were invaluable test subjects for the astronauts who eventually flew in Mercury and subsequent programs.

In these early programs, from Project Mercury to the Space Shuttle, life support systems were primarily open-loop, with few regenerable systems. But with the advent of the International Space Station (ISS), design drivers changed significantly. The ISS would require at least 20 years of continuous operation, which meant engineers had to try to develop ways of increasing the closure of the ISS life support system. How successful were they? Well, we'll get to that in

► Chap. 5. In this chapter, we will focus on the evolution of life support systems (sticking with American systems to make this a little easier), starting with Project Mercury.

Mercury

The launch of Sputnik on 4 October 1957 was the catalyst for creating Project Mercury in 1958. This project included a series of one-man suborbital and LEO missions. NASA selected seven test pilots as astronauts. Each astronaut was between 35 and 40 years old, stood no taller than 180 cm, and was in excellent health. The primary goal of Project Mercury was to place an astronaut in orbit and return him safely. The first orbital flights were planned to last up to 4.5 hours [1]. Life support systems were developed by McDonnell Aircraft Corporation and the Garrett Corporation, which subcontracted to McDonnell. The life support system (LSS) for the Mercury capsule was designed to the following requirements:

(a) Provide metabolic oxygen, pressurization, and ventilation in the pressure suit (◘ Fig. 4.1) and cabin for at least 28 flight hours.

◘ **Fig. 4.1**   Mercury astronauts wore a basic two-layer pressure suit (the Navy Mark IV). Credit: NASA

(b) Maintain a cabin temperature between 50°F and 80°F.
(c) Remove carbon dioxide and water produced by the astronaut.
(d) Maintain comfortable humidity and temperature inside the pressure suit.
(e) Operate in microgravity and in high G conditions.

Mercury Life Support Subsystems

The system maintained a capsule pressure of 5 psi of pure oxygen. As a backup, the astronaut wore a pressure suit. The LSS was operated automatically with a manual control in the event of a malfunction. In essence, the Mercury LSS comprised just two subsystems: the cabin system and the pressure suit system [2]. Oxygen was stored in two spheres pressurized to 7500 psi. Each contained 4 pounds of oxygen, which permitted a flight of up to 26 hours based on a consumption rate of 500 cc per minute and a cabin leakage rate of no more than 300 cc per minute. Each oxygen container was fitted with a filler valve for servicing and a pressure transducer which provided data on the oxygen pressure [3, 4]. Coolant was provided by a water tank that flowed into heat exchangers. Electrical power was a 115–volt 400 cycle AC.

A Regular Atmosphere or Pure Oxygen?

There was a lot of disagreement amongst life support engineers on the subject of capsule atmosphere. Some engineers argued for a normal sea-level pressure while others made the case for a pure oxygen atmosphere. The safer atmosphere was a normal atmosphere because a high oxygen concentration represented a fire hazard and can cause hyperoxia. The problem was that designing a normal atmosphere capsule would have increased design complexity and also increased the risk of hypoxia which would have required sensors to monitor the partial pressure of oxygen. Ultimately, it was decided that the physiological risks of hypoxia were greater than hyperoxia [5].

Pressure Suit

The pressure suit provided breathable oxygen, removed metabolic products, and controlled temperature. The suit was connected to the LSS via an inlet connection at the torso and an exit connection on the helmet (□ Figs. 4.2, 4.3, and 4.4). Oxygen was circulated inside the suit by a compressor. To remove carbon dioxide and waste products, gas was directed through a solid trap to remove particles and then through a chemical canister to remove the carbon dioxide [5, 6]. Finally, the gas was directed through a heat exchanger, which cooled the gas to 45°F – cool enough for water vapor to condense into droplets. These droplets were then directed into a water separator. Suit pressure was maintained by a pressure regulator, which metered oxygen into the suit circuit. In the event of pressure failure, a sensor directed extra oxygen into the suit at a pressure of 0.05 pounds per minute.

4

◻ Fig. 4.2 Each suit was custom-made for each astronaut. The suit, which covered all the body except the head and hands, comprised two layers: an inner gas retention material of neoprene and an outer layer of heat-reflective, aluminized nylon. Credit: NASA

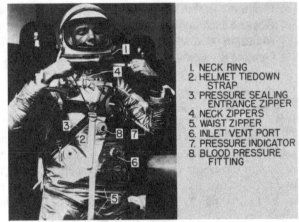

1. NECK RING
2. HELMET TIEDOWN STRAP
3. PRESSURE SEALING ENTRANCE ZIPPER
4. NECK ZIPPERS
5. WAIST ZIPPER
6. INLET VENT PORT
7. PRESSURE INDICATOR
8. BLOOD PRESSURE FITTING

Pressure suit torso.

◻ Fig. 4.3 The ventilation inlet port was located just above the astronaut's waist. The port connected to a manifold inside the suit, from which tubes led to the upper and lower body. Credit: NASA

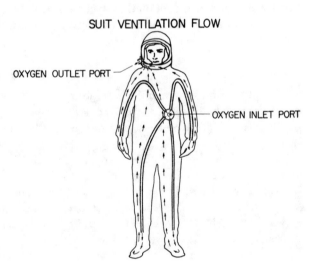

SUIT VENTILATION FLOW

OXYGEN OUTLET PORT

OXYGEN INLET PORT

Cabin Control System

The cabin control system controlled pressure and temperature [1, 2]. The upper limit of cabin pressurization was controlled by a relief valve, which operated automatically. In the event of a fire or a release of toxic gas, the cabin could be manually decompressed using a control handle located on the instrument panel. Oxygen was provided from a 1-pound container pressurized to 7500 psi. A visual indication of the oxygen pressure was provided to the astronaut via light on a sequence panel. Safety valves fitted with pressure sensors ensured that oxygen pressure did not fall below 4 psi. Cabin temperature was maintained by a heat exchanger [1, 2].

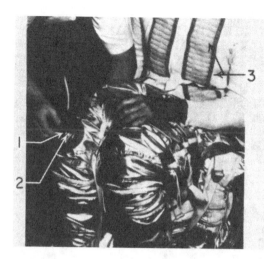

I. INTERNAL PLUG
2. RECEPTACLE PLATE
3. UNDERGARMENT WITH
SPACER PATCHES

Bioconnector (installation).

☐ **Fig. 4.4** The bioconnector. This feature enabled biomedical leads to pass through the suit. Credit: NASA

Instrumentation

The LSS was located in the upper right-hand corner of the instrument panel, which provided information on cabin pressure, humidity, oxygen partial pressure, primary and emergency oxygen supply pressure, and carbon dioxide partial pressure. Next to this panel was a warning light panel that provided auditory and visual warnings in the event of a system failure. Warning lights were provided for loss of pressure, depletion of oxygen, decrease in oxygen partial pressure below 3 psi, increase in carbon dioxide partial pressure above 3 percent in the pressure suit, and excessive cooling water to the suit [1, 2]. On the left of the console, there were controls for cabin decompression and repressurization.

System Operation

During launch, the astronaut was coupled to the pressure suit control system with visor down. Freon was directed into the heat exchangers for cooling. During flight, the cabin was purged by the launch supply, and cabin pressure was normalized at 5 psi. If oxygen partial pressure was adequate, the astronaut could open the visor. In preparation for reentry, the cabin was precooled by opening the heat exchanger water control valves [1, 2]. Following reentry, once a breathable external atmosphere was reached, a snorkel provided ambient air for breathing.

Medical Support

Medical support was divided across three areas of responsibility: medical maintenance of astronauts, preflight and inflight assessment of astronaut health, and postflight evaluation of astronaut response to spaceflight. Predictive values for physiological data during spaceflight were derived from centrifuge runs, flight simulations, and data from flights in high-performance

4

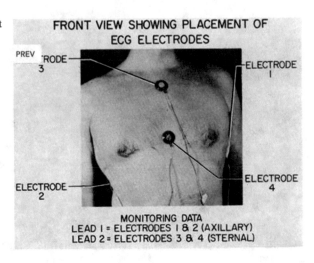

■ **Fig. 4.5**   Electrode placement during Project Mercury. Credit: NASA

aircraft [7]. It was decided that spaceflight required continuous monitoring of physiological data, which was achieved by using a suite of noninvasive bioinstrumentation sensors (■ Fig. 4.5). Body temperature was measured rectally on all missions except the final mission; on this mission, Gordon Cooper used a thermistor [8]. Inflight blood pressure data was not technically feasible during the first four missions, but was measured during the MA-8 and MA-9 flights, which also featured an impedance pneumograph system that measured respiration rates [9, 10, 11]. Electrocardiograph data was obtained, thanks to an electrocardiogram electrode designed specifically for Project Mercury. Body weight was measured preflight and postflight. Measurements were taken nude with empty bladder [8].

In addition to the suite of sensors that provided data inflight, astronauts were subjected to myriad tests postflight which included urine tests, blood chemistry analysis, vital signs (heart rate, respiration, and blood pressure), body mass, body fluid volume, and body weight.

Physiological Measurements

The biomedical assessment of Project Mercury was that humans could function in space for flight durations that exceeded 1 day. The data can be best divided into inflight and postflight (■ Tables 4.1 and 4.2) as follows:

Inflight

Generally, heart rates were higher than those reported in centrifuge runs, while respiration rates were higher during lift-off than in simulations. Blood pressure data was similar to what had been observed in preflight simulations, and changes in blood pressure were consistent with reported weight loss. A general observation

◘ **Table 4.1** Preflight and postflight temperature and heart rate data

Flight	Temperature (°F)			Heart rate (bpm)		
	Preflight	Postflight	Change	Preflight	Postflight	Change
MR-3	99.0	100.2	1.2	68	76	8
MR-4	97.8	100.4	2.6	68	90	22
MA-6	98.2	99.2	1.0	68	76	8
MA-7	97.2	97.6	0.4	60	78	18
MA-8	97.6	99.4	1.8	72	92	20
MA-9	97.4	99.4	2.0	76	86	10

◘ **Table 4.2** Preflight and postflight weight loss

Flight	Weight (kg)			% weight loss
	Preflight	Postflight	Change	
MR-3	76.79	75.70	−1.09	1.42
MR-4	68.27	66.80	−1.47	2.15
MA-6	77.79	75.30	−2.49	3.20
MA-7	69.85	67.10	−2.75	3.94
MA-8	80.19	78.20	−1.99	2.48
MA-9	66.68	63.20	−3.48	5.22

was a decrease in systolic pressure which was aligned with an increase in heart rate. The most marked cardiovascular response to spaceflight was observed following Gordon Cooper's flight; following egress of the spacecraft, the astronaut became presyncopal while standing. In Cooper's case, his pulse pressure narrowed, and his mean arterial pressure fell, which are symptoms of *orthostatic intolerance* (OI). What is orthostatic intolerance? One consequence of being weightless is a fluid shift of up to 2 liters from the lower extremities to the upper extremities due to reduced gravity. And when astronauts return to Earth, all that fluid rushes back from the upper extremities to the lower extremities. This causes astronauts to feel faint and "orthostatically intolerant." All Mercury astronauts lost weight during their flights (◘ Table 4.3), and the amount of weight loss (between 1.1 and 3.5 kg or between

4

■ Table 4.3   Project Mercury missions

Mission	Date	Flight duration (h/min/s)	Weightless time (h/min/s)	# Earth orbits	Pilot
MR-3	5/5/1961	15:28	5:04	0	A.B. Shepard
MR-4	7/21/1961	15:37	5:00	0	V.I. Grissom
MA-6	2/20/1962	4:55:23	4:38:00	3	J.H. Glenn
MA-7	5/24/1962	4:56:05	4:39:00	3	M.S. Carpenter
MA-8	10/3/1962	9:13:11	8:56:22	6	W.M. Schirra
MA-9	5/15/1963	34:19:49	34:03:30	22	L.G. Cooper

1.4 and 5.2 percent of body weight) was proportional to the length of the flight. Urine volume, which was measured in MR-4, MA-6, MA-7, and MA-9, revealed an excretion rate of between 30 mL per hour for the MA-9 pilot and 155 mL per hour for the MA-7 pilot, these amounts being consistent with fluid intake [9, 10, 11, 12].

When the six Mercury astronauts ventured into the unknown and dangerous environment of space, no established physiological values for spaceflight existed. Neither did proven methods for determining threshold tolerances. But Project Mercury demonstrated that humans could function in space without significant deterioration of physiological function. The Mercury missions were short but provided flight surgeons and life support engineers with a wealth of data for the following program: Project Gemini.

Gemini

The Gemini LSS, which provided life support for two astronauts, was divided into four subsystems: an oxygen supply subsystem, a water management subsystem, a cooling subsystem, and a suit loop. This LSS provided oxygen for the pressure suits and for cabin pressurization, it removed carbon dioxide and moisture from the suit and the cabin, and it provided for the storage and disposal of water [13].

Primary Oxygen Subsystem

This subsystem stored and provided oxygen for breathing for the pressure suit and cabin. Oxygen pressure in the cabin was maintained by a cabin pressure relief valve. The oxygen capacity for a 2-day mission was 15.3 pounds. Stored cryogenically in a spherical container, the oxygen was heated to a gas using a heat exchanger. Carbon dioxide was adsorbed using a cartridge that removed odors and up to 11 pounds of carbon dioxide. In the event of a decompression, the oxygen supply was turned off automatically when the pressure reached 4 psi and the pressure suit maintained

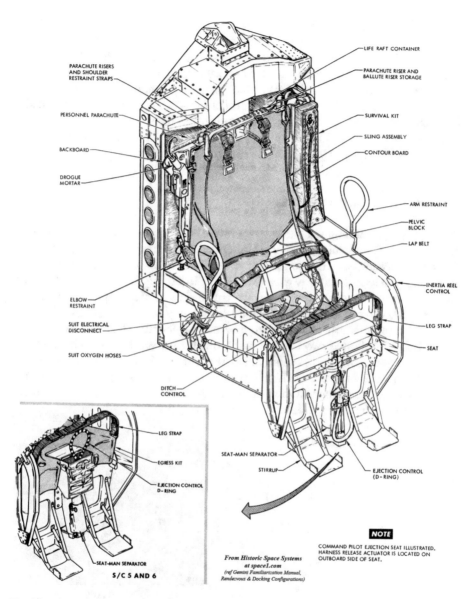

LIFE RAFT CONTAINER

PARACHUTE RISERS
AND SHOULDER
RESTRAINT STRAPS

PARACHUTE RISER AND
BALLUTE RISER STORAGE

PERSONNEL PARACHUTE

SURVIVAL KIT

SLING ASSEMBLY

BACKBOARD

CONTOUR BOARD

DROGUE
MORTAR

ARM RESTRAINT

PELVIC
BLOCK

LAP BELT

INERTIA REEL
CONTROL

ELBOW
RESTRAINT

SUIT ELECTRICAL
DISCONNECT

LEG STRAP

SEAT

SUIT OXYGEN HOSES

DITCH
CONTROL

LEG STRAP

SEAT-MAN SEPARATOR

EGRESS KIT

STIRRUP

EJECTION CONTROL
(D-RING)

EJECTION CONTROL
D-RING

SEAT-MAN SEPARATOR

S/C 5 AND 6

*From Historic Space Systems
at space1.com
(ref Gemini Familiarization Manual,
Rendezvous & Docking Configurations)*

NOTE

COMMAND PILOT EJECTION SEAT ILLUSTRATED,
HARNESS RELEASE ACTUATOR IS LOCATED ON
OUTBOARD SIDE OF SEAT.

◘ Fig. 4.6   Gemini ejection seat assembly. Credit: NASA

pressure [13]. The secondary oxygen subsystem, which comprised two tanks, provided an oxygen flow of 0.08 pounds per minute to each astronaut. The subsystem was triggered when pressure in the primary oxygen subsystem fell below 75 psi. A third oxygen subsystem was the Egress Oxygen Subsystem, which provided oxygen for breathing in the event of an ejection (the Gemini spacecraft was one of the only two spacecraft fitted with ejection seats (◘ Fig. 4.6), the other being the Columbia Space Shuttle). This subsystem contained one-third of a pound of oxygen and was activated manually following ejection.

Water Management Subsystem

This subsystem collected and stored water for drinking and cooling. It comprised water tanks, urine receptacle, drinking nozzle, reservoir, evaporator, and water pressure regulator. One water tank, which held 16 pounds of water, was located in the equipment section of the spacecraft (◘ Fig. 4.7). A second tank, which also contained 16 pounds of water, was located in the reentry module, and another 7 pounds of water was located in the heat exchanger reservoir. Water was forced through the subsystem by diaphragms pressurized by oxygen.

Temperature Control Subsystem

This subsystem maintained cabin temperature, suit temperature, and equipment temperature. The Gemini spacecraft generated three times as much heat as the Mercury spacecraft and did so for almost ten times as long. Because of this, engineers had to devise a better way to reject heat. To do this, they developed a space radiator which effectively comprised the entire skin of the adapter module. The system worked by pumping coolant fluid (a silicon ester fluid – Monsanto MCS 198) from a reservoir, through coolant lines and regenerative heat exchangers. The coolant circuit followed two parallel paths, one for the suit and one for the cabin. Temperature-sensitive valves maintained the outlet coolant between 36°F and 42°F. When the temperature fell below 36°F, coolant was simply directed to the regenerative heat exchanger, and when the temperature exceeded 42°F, the coolant was directed to the space radiator.

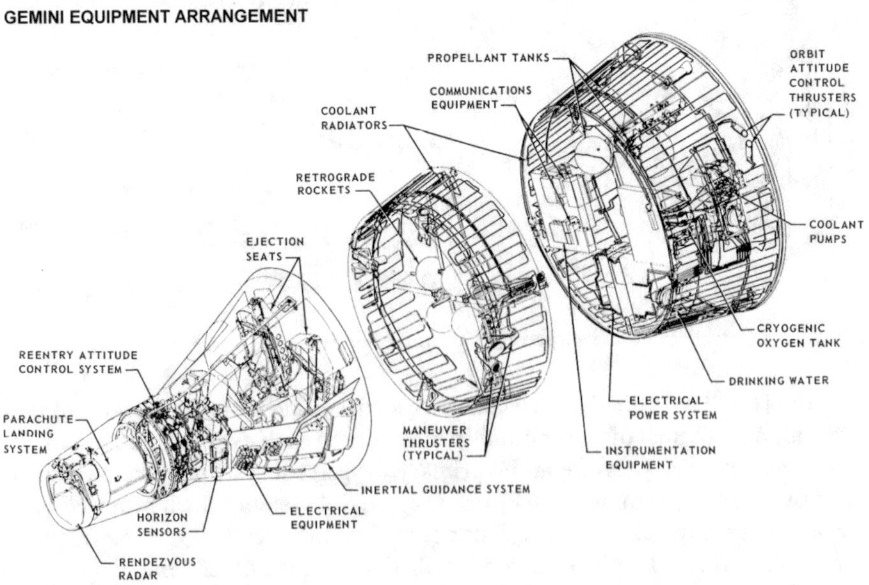

◘ **Fig. 4.7** Arrangement and location of equipment in the Gemini spacecraft. Credit: NASA

Suit Loop

Astronauts were provided with a redundant atmosphere, thanks to a closed-pressure suit circuit that provided cooling, pressurization, purification, and removal of water. The suit loop circulated oxygen through the suit ( Fig. 4.8) and removed carbon dioxide using charcoal and lithium hydroxide. The suit loop had two modes of operation, normal circulation and high-rate mode, which pumped oxygen directly into the suit.

The suit loop, which was a major improvement on the Mercury system, comprised two suit pressure demand regulator valves, four check valves, two throttle valves, a system shutoff, two compressors, a heat exchanger, and a carbon dioxide absorber. In operation, the demand regulator maintained suit pressure at 3.7 psi unless suit pressure was lost, in which case the suit circuit switched to *high rate of operation*, which enabled an oxygen flow rate of 0.08 pounds of oxygen per astronaut until suit pressure was restored. Once suit circuit pressure was restored, the high rate of operation shut-off valve was reset manually.

Fig. 4.8 Gemini III astronauts, Gus Grissom and John Young wearing their David Clark G3C suits. The G3C was the first of three suits (Table 4.4) used in the Gemini Program. Credit: NASA

4

	Gemini G3C space suit	Gemini G4C space suit	Gemini G5C space suit
◘ Table 4.4	Gemini Program space suits		
Missions	Gemini 3, 6, and 8	Gemini 4–6, 8–12	Gemini 7
Ejection	✓	✓	✓
Intravehicular activity (IVA)	✓	✓	✓
Extravehicular activity (EVA)	✗	✓	✗
Suit weight	10.7 kg	15.4 kg	17.3 kg

Storage of Cryogenic Liquids

Oxygen, which was stored in a cryogenic state on board the Gemini spacecraft, comprised a subsystem that weighed about four-tenths a pound for each pound of oxygen stored. This was a significant improvement over the pressurized system used in Mercury.

Physiological Measures

During the Gemini Program, emphasis was placed on the performance of the cardiovascular and musculoskeletal systems. Analysis of this data revealed no serious performance issues related to these areas or in areas such as vision or disorientation (◘ Table 4.5).

One notable observation was the time course of decrements. Peak decrements observed in the 8-day flight were significantly lower in the 14-day flight, which indicated astronauts were adapting to the microgravity environment. Having said that, it was not possible to rule out mission variables such as diet and fluid intake as having an impact on this adaptation. Anomalies in cardiovascular function were observed, as expected, but these variations were considered within the envelope of normality when the effects of acceleration and weightlessness were considered.

EVA

The major concern of biomedical scientists during the Gemini Program was astronaut health in the performance of extravehicular activities. From the biomedical perspective, the success of a spacewalk really came down to the ability of the suit to provide oxygen, maintain suit pressure, maintain temperature and humidity, and at the same time provide adequate body–joint mobility, flexibility, and dexterity. The Gemini EVA suit (◘ Fig. 4.9) comprised a multilayer fabric system that included a comfort liner, a gas bladder, a structural restraint, and an outer protective layer.

▣ **Table 4.5** Predictions and observations of human response to spaceflight during the Gemini Program [14]

Predicted	Observed	Predicted	Observed
Electromechanical delay in cardiac cycle	None	Stimulant need	Occasionally before reentry
Reduced exercise capacity	None	Infectious disease	None
		Fatigue	Minimal
Reduced blood volume	Moderate	Circadian rhythms	No disruption
Reduced plasma volume	Minimal		
Dehydration	Minimal	Skin infections and breakdown	Dryness, including dandruff
Weight loss	Variable		
Bone loss	Minimal	Sleeplessness	Minor
Loss of appetite	Minimal		
Nausea	None	Reduced visual acuity	None
Muscular incoordination	None	Disorientation and motion sickness	None
Muscular atrophy	None		
Hallucinations	None		
Euphoria	None		
Psychomotor performance	Not impaired	Cardiac arrhythmias	None
		High blood pressure	None
Sedative need	None	Low blood pressure	None
		Fainting postflight	None

Inside the suit, oxygen was directed for consumption and thermal control using a gas distribution system. Environmental control of the suit's LSS was provided via an extravehicular LSS that comprised a chest pack, together with hoses and connectors for inlet and output of gases. On the whole, the suit performed well, although there was a tendency for astronauts to become overheated, and astronauts complained of the equipment packages being too bulky. Another problem was exhaustion, which was reported by astronauts conducting spacewalks during the Gemini 9A and Gemini 11 flights. A part of the reason the astronauts became fatigued was attributed to lack of sleep, lack of adequate training, and exhaustive preflight training [14].

The Gemini flights demonstrated the successful operation of the life support systems for flights approaching 14 days. No significant problems were identified in any of the subsystems, and astronauts reported no problems when drinking, eating, or performing bodily functions. Data also revealed there were no indications that contaminants or radiation reached significant levels. These findings, which are summarised in NASA document SP-4213 (▶ Chap. 7), provided LSS engineers with a solid foundation to develop the LSS of the subsequent program: Apollo.

4

Fig. 4.9 Gemini 4. Ed White making the first spacewalk by an American on 3 June 1965. The EVA lasted 23 minutes. Credit: NASA (the photo was taken by Gemini 4 Commander, Jim McDivitt)

Apollo

The Apollo environment control system comprised two environment control systems (ECS): one for the Command Module and one for the Lunar Module.

Command Module ECS

This life support system (**Fig. 4.10**) functions can be summarized as follows:
1. Oxygen atmosphere maintained at 5 psia
2. Astronauts to work in shirt-sleeve mode except for critical mission phases
3. Cabin pressure maintained at 3.5 psia under defined emergency conditions
4. Carbon dioxide (CO_2) removal by lithium hydroxide (LiOH) absorption and limited to a partial pressure of 7.6 mm Hg
5. Cabin temperature maintained at 75° ± 5°F with relative humidity limited to 40 to 70 percent
6. Thermal control provided for the electrical and electronic equipment

To accomplish these functions, the ECS comprised six subsystems as follows:
1. Oxygen
2. Pressure suit circuit
3. Water
4. Coolant
5. Waste management
6. Postlanding ventilation

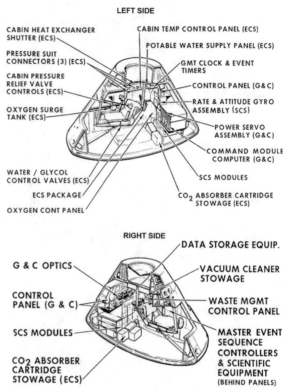

□ **Fig. 4.10** Schematic showing features of the Apollo Command Module's ECS. Credit: NASA

APOLLO COMMAND MODULE INTERIOR

LEFT SIDE

CABIN HEAT EXCHANGER SHUTTER (ECS)
PRESSURE SUIT CONNECTORS (3) (ECS)
CABIN PRESSURE RELIEF VALVE CONTROLS (ECS)
OXYGEN SURGE TANK (ECS)
WATER / GLYCOL CONTROL VALVES (ECS)
ECS PACKAGE
OXYGEN CONT PANEL

CABIN TEMP CONTROL PANEL (ECS)
POTABLE WATER SUPPLY PANEL (ECS)
GMT CLOCK & EVENT TIMERS
CONTROL PANEL (G&C)
RATE & ATTITUDE GYRO ASSEMBLY (SCS)
POWER SERVO ASSEMBLY (G&C)
COMMAND MODULE COMPUTER (G&C)
SCS MODULES
CO₂ ABSORBER CARTRIDGE STOWAGE (ECS)

RIGHT SIDE

DATA STORAGE EQUIP.
G & C OPTICS
VACUUM CLEANER STOWAGE
CONTROL PANEL (G & C)
WASTE MGMT CONTROL PANEL
SCS MODULES
MASTER EVENT SEQUENCE CONTROLLERS & SCIENTIFIC EQUIPMENT (BEHIND PANELS)
CO₂ ABSORBER CARTRIDGE STOWAGE (ECS)

Oxygen Subsystem

This subsystem, which controlled the distribution of oxygen inside the Command Module (CM), was supplied from the Service Module (SM) cryogenic tanks [15]. It had the following functions:

1. Storage of a reserve oxygen supply
2. Regulation of oxygen pressure inside the CM
3. Controlling CM cabin pressure
4. Controlling CM cabin pressure in emergency mode
5. Purging of the pressure suit circuit

The oxygen capacity of this subsystem was 78 kilograms for a 14-day mission (820 grams per astronaut for consumption and 2.18 kilograms for cabin leakage, with additional allowance for EVAs), although the actual usage was lower because of lower than anticipated cabin leakage (□ Tables 4.6 and 4.7). This subsystem performed well across all the Apollo missions, as evidenced by the fact that no emergency cabin pressure regulation was required and all planned depressurizations and repressurizations were completed successfully [15].

4

◘ Table 4.6   Apollo Command Module oxygen consumption [16]

Item	Specification requirement (14 days)		Apollo 15 mission (12.3 days)	
	Kg	**(lb)**	**kg**	**(lb)**
Crew consumption	34.29	(75.6)	22.09	(48.7)
Cabin leakage	30.48	(67.2)	2.68	(5.9)
Cabin repressurizations	5.31	(11.7)	4.08	(9.0)
One CM puncture	1.63	(3.6)	-	.
LM support	6.58	(14.5)	5.94	(13.1)
Tank bleeds	.	.	4.45	(9.8)
Cabin and WMS purges	.	.	3.49	(7.7)
EVA flow	.	.	6.67	(14.7)
Totals	78.29	(172.6)	49.40	(108.9)

◘ Table 4.7   Actual ECS oxygen consumption across all missions [16]

Apollo mission	Duration days/hours	Oxygen consumed kg
7	10:20	46.26
8	6:03	23.13
9	10:01	44.91
10	8:00	32.21
11	8:03	37.19
12	10:05	44.91
13	5:23	13.61
14	9:00	42.64*
15	12:07	49.44**
16	11:02	48.08**
17	12:14	49.90**

*Includes 4.5 kg for high flow demonstration test of cryogenic system
**Includes 11 to 13 kg for EVA flow and cabin repressurization

Pressure Suit Circuit

This second subsystem provided the following functions:
1. Provided the astronauts with a conditioned atmosphere
2. Controlled suit gas circulation automatically
3. Controlled temperature
4. Removed debris
5. Removed moisture and odors
6. Removed carbon dioxide from the cabin and the suit

This subsystem performed well across all missions. Pressure regulation was maintained within the required 3.5–4 psia range. The atmosphere in the CM was 60 percent oxygen and 40 percent nitrogen mix with the suit circuit at 100 percent oxygen. Pressure sensors indicated suit-to-cabin differentials, and a valve was used to ensure a constant flow of 0.23 to 0.32 kilograms of oxygen per hour. Carbon dioxide was removed by lithium hydroxide absorber elements capable of removing carbon dioxide at a rate of 0.064 kilograms per hour for 24 hours. Two of these systems operated in parallel and maintained the carbon dioxide partial pressure below 7.6 mm Hg. When all three astronauts were living in the spacecraft, the elements were changed every 24 hours [15].

Water Subsystem

The water subsystem:
1. Received potable water, which was produced as a by-product of operating the fuel cells
2. Stored water
3. Chilled and heated the water
4. Included a wastewater section which collected and stored water that was extracted from the suit heat exchanger
5. Dumped water that was excess to system requirements

This subsystem managed between 180 and 225 kilograms of water. The fuel cell production rate of 0.68 to 0.91 kilograms per hour far exceeded crew requirements, which led to most of the water being dumped overboard. The water balance for Apollo 15 is presented in ◘ Table 4.8.

One improvement the astronauts appreciated was hot water, although this particular function was the source of some negative comments from the crew because the system had a habit of introducing gas into the water. One cause of the gas was fuel cell operation, because water produced by the system was saturated with hydrogen gas. On Apollo 12, and subsequent missions, this gas was removed by passing the water through a hydrogen separator. Another source was oxygen in the bladder for storing water. This bladder had to be pressurized to expel water, but oxygen permeated the bladder material with the result that bubbles formed in the food bags. This problem was solved in time for the Apollo 11 mission by installing a gas separator cartridge. Water sterility was achieved using a chlorine solution contained in Teflon ampoules [15].

4

◘ Table 4.8 ECS water balance for Apollo 15 [16]	
Initial onboard water	Kilograms
Potable water	13.15
Wastewater	12.25
Subtotals	25.40
Fuel cell production	235.57
LiOH reaction	12.25
Metabolic oxidation	11.79
Subtotals	259.91
Potable tank	14.06
Waste tank	23.13
Body wastewater	43.09
Evaporator operation	3.63
Waste tank	191.42
Potable tank	7.26
URA flushing and samples	2.72
Subtotals	248.12

Coolant Subsystem

The coolant subsystem:
1. Comprised a water/ethylene glycol coolant
2. Supplied cooling for the pressure suit circuit
3. Provided a potable water chiller
4. Supplied heating and cooling for the cabin atmosphere
5. Provided a heat rejection mechanism via primary and secondary coolant loops

This subsystem maintained adequate thermal control throughout all missions. Part of this control was achieved using a slow, controlled roll of the CM (also called the barbecue maneuver) that ensured the radiator outlet temperature rarely exceeded 10° C. The coldest coolant was passed through the suit heat exchanger to ensure gas cooling and removal of condensate. Typical heat load and rejection required between 1170 and 1470 watts.

Waste Management

The waste management subsystem:
1. Permitted the dumping overboard of urine
2. Stored and vented solid waste

Waste management was and still is one of the most bothersome challenges of manned spaceflight. Before Apollo 12, astronauts used a urine transfer system (UTS) comprising a rubber cuff connected to a flexible collection bag. Following Apollo 12, a urine collection and receptacle assembly (UCTA) was used, as depicted in ◘ Fig. 4.11. In addition to sampling myriad physiological metrics, flight surgeons also sampled urine output (◘ Table 4.9).

When astronauts wore space suits during EVAs, the UCTA was worn (◘ Fig. 4.12) over the liquid cooling garment. When it came to dealing with solid waste, astronauts relied on a plastic bag that was taped (using Stomaseal tape) to the nether regions to capture waste. After completing his ablutions, the astronaut sealed the bag and kneaded it to mix bactericide with the contents. During surface operations, the bag fecal collection system was not feasible, so a pair of astronaut diapers were used [15].

From an engineering perspective, the Apollo waste management system worked fairly well, but astronauts didn't give the system many favorable reviews. One recurring problem was the amount of manipulation required to operate the system, and another issue was the frequency of urine spills caused by the challenges of manipulation. More challenges were presented when using the fecal bags, which required unique skillsets to prevent accidental release of the bag's contents! Another complaint leveled at fecal bag operation was the amount of time required – 45 minutes on a good day – to accomplish the process [15]. Following several complaints from Apollo astronauts, attempts were made to upgrade the system for the Apollo 16 crew, but these attempts were judged by the crew of that mission to have failed.

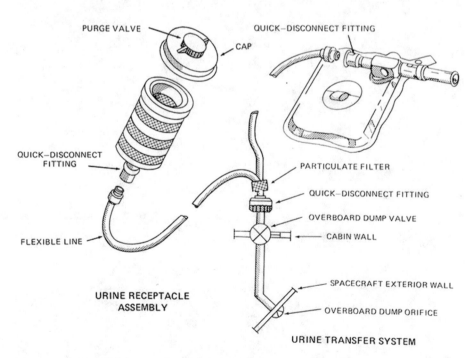

◘ **Fig. 4.11** The urine receptacle assembly. Credit: NASA

4

■ **Table 4.9**   Apollo 17 urine sampling data [16]

Crewman	Time of sampling (ground elapsed time (GET))			
	Preflight, predicted (hr/min)	Actual (hr/min)	Sample volume (ml)	Calculated pooling volume (ml)
CMP	18:30	18:50	110.7	1154
	35:00	34:36	85.5	811
	58:45	58:22	91.0	1875
	83:30	83:22	89.9	1034
	107:00	110:00	83.2	1500
	133:00	133:00	86.3	769
	156:10	156:10	74.8	1667
	180:45	180:40	104.9	2000
	208:00	208:30	70.4	1500
	230:25	230:28	84.0	1200
	252:50	252:45	93.7	1304
	276:50	276:30	89.8	938
	300:30	299:50	116.1	1667
LMP	18:30	18:30	84.8	750
	35:00	34:40	78.8	448
	58:45	58:20	118.0	789
	83:30	83:20	74.8	789
	107:00	110:00	78.8	1250
	230:25	230:30	71.9	714
	252:50	252:15	80.9	1111
	276:50	276:25	87.1	1304
	300:30	300:15	104.7	1579
CDR	18:30	18:46	82.0	395
	35:00	34:40	38.7	337
	58:45	58:10	94.0	750
	83:30	83:15	60.1	652
	107:00	110:00	71.1	938
	230:25	230:28	90.2	1000
	252:50	252:50	98.9	1429
	276:50	276:30	108.6	1154
	300:30	299:52	137.3	2500

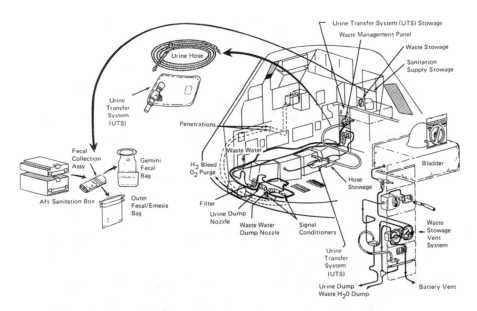

Fig. 4.12 The UCTA. Credit: NASA

Lunar Module Environmental Control System

The Lunar Module (LM) ECS comprised four subsystems:
1. Atmosphere revitalization
2. Oxygen supply and cabin pressure control
3. Water management
4. Heat transport

Atmosphere Revitalization System (ARS)

This subsystem comprised a suit circuit and a suit liquid cooling assembly. The suit circuit assembly, which comprised a closed-loop system that cooled and provided ventilation for the pressure garment assemblies, worked in tandem with the suit liquid cooling assembly that circulated water through the liquid cooling garment and also helped remove dust from the LM's cabin [17].

Oxygen Supply and Cabin Pressure Control Section (OSCPCS)

The OSCPCS (☐ Fig. 4.13) stored oxygen and supplied oxygen to the suit circuit, the cabin, and also the portable life support system (PLSS). This subsystem was also responsible for maintaining cabin pressure [17]. Following transposition and docking, the LM was pressurized and pressure decay was monitored. Between Apollo 11 and 17, the leak rate was between 0.03 and 0.05 pounds per hour (the maximum allowable leak rate was 0.2 pounds per hour). The amount of oxygen consumed during the Apollo 17 mission was 46.2 pounds (the flight prediction had calculated 45.5 pounds).

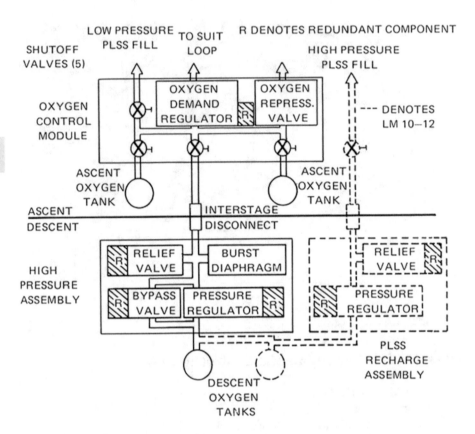

o Fig. 4.13 The OSCPCS. Credit: NASA

Water Management System (WMS)

The WMS supplied potable water for the crew and also refilling of the PLSS water tank. Another function of this system was cooling the pressure garment assemblies, the cabin, and the electronic equipment. It achieved this function using heat transport section sublimators, which comprised coolant loops through which water and ethylene glycol were routed. Typical water consumption for a mission was about 400 pounds, of which 2.3 pounds was used for filling the sublimator, 10 pounds was used for refilling the PLSS, and 8 pounds was used for metabolic waste [17]. The process of dealing with waste on the LM was different than on the CM because there was no dumping of waste on the lunar surface (imagine the media blitz if that had happened!). Instead, the astronaut relied on in-suit urine containers and a urine transfer hose to drain the urine into a waste fluid container.

Life Support System Issues

It was Apollo 13 that perhaps more than any other mission put the spotlight on the limitations and versatility that was the Apollo Program's LSS. Apollo 13 was aborted 56 hours after launch, when the oxygen supply in the Service Module

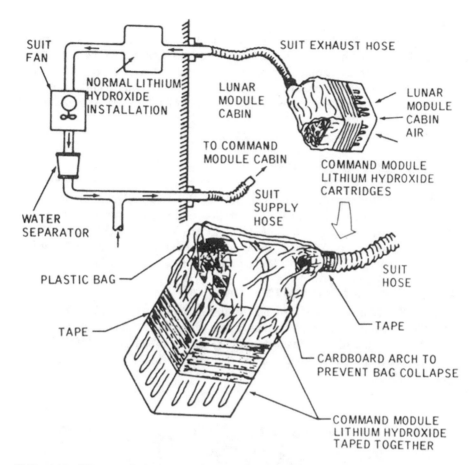

◘ Fig. 4.14 "Houston, we have a solution!" How the Apollo 13 crew solved the carbon dioxide removal problem – actually, it was the engineers in the Mission Evaluation Room that solved the problem and then sent up the solution to the astronauts. Credit: NASA

(SM) was lost and the CM's ECS no longer had a supply of oxygen, water, or electrical power. To preserve what little life support consumables the crew had, the repressurization package tanks were isolated, the water tanks were depressurized, and the CM was powered down. The crew moved into the LM, which served as a lifeboat for the remaining 83 hours (twice the intended limits of the LM) of the mission. Once the transfer to the LM had been completed, issues started to arise immediately. Power levels had to be conserved to limit heat loads, and water consumption rate had to be limited to reduce thermal loading. Since there were not enough lithium hydroxide canisters in the LM, the crew had to cobble together and engineer a system (◘ Fig. 4.14) to scrub carbon dioxide using space suit return hoses taped to plenum chambers (using duct tape - any drama always includes duct tape!). This system was used for almost 2 days until the CM was reactivated and the LM jettisoned.

4

Dust

Dust (◘ Fig. 4.15) was a major life support concern (other concerns are listed in ◘ Table 4.10) throughout Apollo surface operations. Despite the best efforts of the LM astronauts, contamination of the CM occurred after every surface stay. Filters were developed to speed up the capture of lunar contaminants, handheld vacuum cleaners were used, and continuous cycling of cabin gas was tried, but all to no avail.

Extravehicular Mobility Unit

The Apollo Extravehicular Mobility Unit (EMU) was designed for a unique set of tasks, one of which was the exploration of the lunar surface. The EMU (◘ Fig. 4.16), which comprised a pressure garment assembly (PGA) and a PLSS, enabled traverses of up to 7 hours to be made across the lunar surface. What follows here is a description of the EMU used on the Apollo 11 mission.

Pressure Garment Assembly

The PGA (◘ Table 4.11) was designed as an IVA configuration, which was worn by the CM pilot, and as an EVA configuration, which was worn by the commander and LM pilot [15]. Both versions comprised a torso–limb suit assembly (TLSA), an over layer, a helmet, gloves, controls, and an instrumentation and communication equipment (◘ Fig. 4.17).

◘ **Fig. 4.15** Last man on the Moon, Gene Cernan, covered in lunar regolith. Credit: NASA

■ Table 4.10 Apollo mission problem summary [16]

Problem description	Apollo mission	Cause	Mission impact	Corrective action	Recommendation for future systems design
Oxygen subsystem					
High oxygen flow (procedural error)	Most	Manual overboard dump valve remained open	Increased O2 usage	None	Include time-to-close feature in manual purge valves
Slow oxygen tank repressurization	9	Valve indicator misaligned – valve partially closed	None	Preflight inspection	Include greater detent identification or integral position indicators
Cabin fans					
(a) Noisy	All	Lack of noise suppression	Discontinued most fan use	None	Add acoustical design requirements
Failed to operate	9	Foreign objects in fan area	.	Inspection	Protect fan inlet and outlet with screens
Discrepant CM/LM [delta] gage readings	15	Valve position arrow chipped off	Confusion during integrity check	Metalcal arrow substituted	Design indicators into manual devices
Pressure suit circuit subsystem					
Return filter screen partially plugged	All	Cabin debris from manner operations	Required daily crew cleaning	Incorporated in crew procedures	Design filters for accessible cleaning or replacement
Free water in suit hoses	7; 15	Prelaunch degradation of suit heat exchanger condensate flow	None (droplets minor)	Improved servicing techniques	Minimize the use of sintered plates or design for in place restoration to original

(continued)

4

□ **Table 4.10** (continued)

Problem description	Apollo mission	Cause	Mission impact	Corrective action	Recommendation for future systems design
LiOH elements sticking in ECU canister	16	Adverse operating conditions – temp control failure and flow valves mispositioned	None	Preflight fit check requirements tightened, crew procedures revised	Design for contingency conditions, use automatic valves to minimize crew operation
Coolant subsystem					
Evaporator dry out	7, 8, 9	Wick sensor location not representative of wick wetness	Manual reservicing required	Removed sponges locally near sensors	Adequately develop test liquid systems in six axes to verify operation in zero gravity
Evaporator dry out	10	Micro switch maladjustment	Manual reservicing required	Added preflight inspection	Limit switches should be individually verified after installation
Condensation on cold coolant lines	All	Lines not fully insulated	None	Increased line insulation	Provide adequate insulation for lines operating below dew point. Locate coldest lines away from electronics if possible
Primary accumulator quantity decayed	11	Valve not fully closed due to excessive knob play	None	None	Provide greater detents, eliminate all play between knob and valve
Glycol temperature exceeded control tolerance	11	Bearing failure in control valve drive mechanism	None (recovered satisfactorily)	None	Use limit switches on control valves to prevent continuous drive signals when valves are on end stops

Coolant subsystem (continued)

Higher than expected radiator outlet temperatures	15	Lunar attitude holds and possible radiator coating degradation during launch	Increased cabin condensation am excessive temperature excursions	None (launch procedures unique to Apollo 15)	Protect thermal coatings from mission contamination if possible. Configure for minimum attitude hold impact
Glycol temperature controller failed in automatic mode	16	Silicon controlled rectifiers (SCR) turned on without a gate drive signal	Manual control required with increased condensation and LiOH element swelling	Controllers screened to determine condition of SCRs	Include proper part derating and part application in electronics design phase
Water subsystem					
Gas in potable water	All	Fuel cell carrier gas and bladder permeability	Crew discomfort	Added hydrogen separator and gas separator cartridge assemblies	Eliminate bladder and gas blanket type tanks, provide adequate gas separators
Hot water valve leakage	12, 14	Tolerance buildup with heated water	None	Added hot water expulsion tests	Provide system checkout in all operational modes
Potable water tank failed to fill	15	Contamination on check valve seat	None	None	Verify system cleanliness and filter all fluids entering the spacecraft

(continued)

□ Table 4.10 (continued)

Problem description	Apollo mission	Cause	Mission impact	Corrective action	Recommendation for future systems design
Water subsystem (continued)					
Chlorine and buffer ampoules leaked when injected	15, 16, 17	Inner bag breakage due to bonding problems and pinching between wall and end plate	Required additional crew time and cleanup	Added inspection requirements and revised crew procedures	Provide automatic or semiautomatic systems to reduce crew operation
Waste management subsystem					
Urine filters partially clogged	12	Urine breakdown due to overnight storage	Required replacement with spares	Revised crew procedures; added larger filter for mission requiring storage	Anticipate contingency requirements during initial design
Urine dump line partially frozen	14	Undetermined	Urine backup – temporary blockage	Revised crew procedures to minimize flush and require gas purge	Minimize lengths of dump lines, provide adequate heater and orifice sizes
Miscellaneous					
Vacuum cleaner failed to operate	16	Dust accumulation between turbine wheel and housing	Cleanup time lengthened	Revised crew procedures	Adequately protect all fans with filters
Instrumentation calibration shifts, erratic operations, and failures	Many	Contamination, internally and externally generated, corrosion, electronic failures	Backup instrumentation utilized, flight procedures modified	Internal epoxy coating added, inlet filters provided, corrosion displaced by cycling, modified design	Protect pressure sensors with filters including static sense lines, since contamination migration may occur in zero gravity

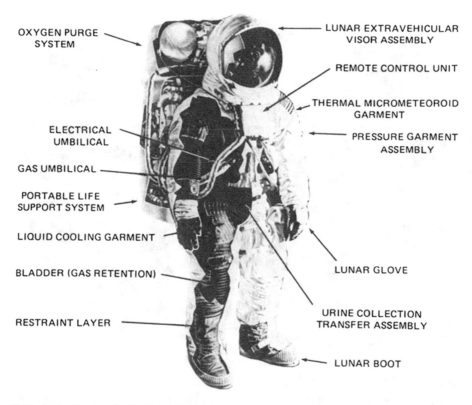

□ Fig. 4.16 The Apollo EMU. Credit: NASA

□ Table 4.11	PGA characteristics
Weight	19.69 kg
Operational temperature limitations	± 394°K
Leak rate at 25 511 N/m² (3.7 psig) (max)	180 scc/min
Operating pressure	25 855 ± 1724 N/m²
Structural pressure	41 369 N/m²
Proof pressure	55 158 N/m²
Burst pressure	68 948 N/m²

4

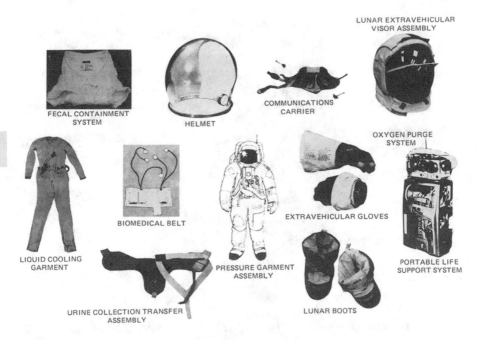

Fig. 4.17   PGA features. Credit: NASA

Torso–Limb Suit Assembly

The torso–limb suit assembly (TLSA) comprised the part of the PGA that covered the body with the exception of the head and hands. The pressure bladder of the TLSA (■ Fig. 4.18) was a neoprene-coated nylon fabric covered by a nylon restraint layer. Built into the right thigh section was a biomedical injection port to enable self-administration of a hypodermic injection without compromising gas retention. The PGA's inner layer was nylon, to which were attached a series of non-collapsible ducts designed to improve ventilation. The ventilation system directed gas flow to the helmet to help the astronaut breathe and also to improve defogging [16].

Pressure Helmet Assembly

The pressure helmet assembly (PHA) featured a visor that was made of optical quality polycarbonate plastic (■ Fig. 4.19). Within the PHA was a feed port that permitted the insertion of a probe for water and food and at the back of the assembly was a foam vent pad that acted as a headrest. The lunar visor assembly comprised three eyeshades and two visors, the outer of which was the sun visor, which was made of high temperature polysulfone plastic. Together with the inner visor, the sun visor (■ Fig. 4.20) helped protect the astronaut from infrared rays and also micrometeoroid damage.

Pressure Gloves

The pressure glove, which was a multilayered assembly, was attached to the TLSA by a quick disconnect coupling. Abrasion protection was provided by Chromel-R, and the thumb and fingertips were made of high-strength rubber-coated nylon.

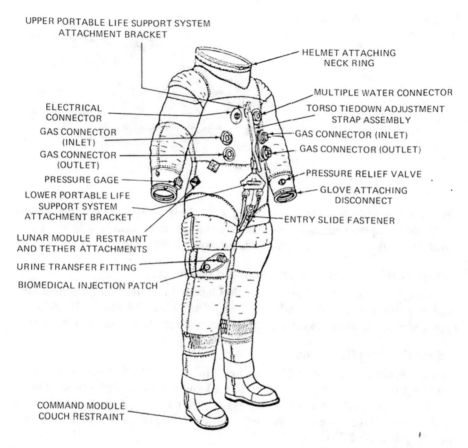

UPPER PORTABLE LIFE SUPPORT SYSTEM
ATTACHMENT BRACKET

HELMET ATTACHING
NECK RING

MULTIPLE WATER CONNECTOR

ELECTRICAL
CONNECTOR

TORSO TIEDOWN ADJUSTMENT
STRAP ASSEMBLY

GAS CONNECTOR
(INLET)

GAS CONNECTOR (INLET)

GAS CONNECTOR
(OUTLET)

GAS CONNECTOR (OUTLET)

PRESSURE GAGE

PRESSURE RELIEF VALVE

LOWER PORTABLE LIFE
SUPPORT SYSTEM
ATTACHMENT BRACKET

GLOVE ATTACHING
DISCONNECT

ENTRY SLIDE FASTENER

LUNAR MODULE RESTRAINT
AND TETHER ATTACHMENTS

URINE TRANSFER FITTING

BIOMEDICAL INJECTION PATCH

COMMAND MODULE
COUCH RESTRAINT

◘ Fig. 4.18 Extravehicular configuration of torso limb suit assembly. Credit: NASA (Publication: SP-368 Biomedical Results of Apollo)

◘ Fig. 4.19 The Apollo pressure helmet. Credit: NASA

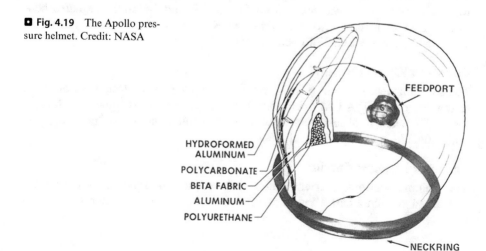

FEEDPORT

HYDROFORMED
ALUMINUM

POLYCARBONATE

BETA FABRIC

ALUMINUM

POLYURETHANE

NECKRING

4

◘ Fig. 4.20 The Apollo lunar extravehicular visor assembly (LEVA). Credit: NASA

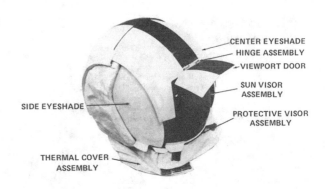

CENTER EYESHADE
HINGE ASSEMBLY
VIEWPORT DOOR
SUN VISOR ASSEMBLY
PROTECTIVE VISOR ASSEMBLY
SIDE EYESHADE
THERMAL COVER ASSEMBLY

The shell assembly was fitted to the pressure glove using fastener tape, anchor straps, and neoprene adhesive. For comfort and to aid in sweat absorption, astronauts wore nylon tricot gloves under the pressure gloves [16].

Integrated Thermal Micrometeoroid Garment

The Integrated Thermal Micrometeoroid Garment (ITMG) comprised a multi-laminate assembly that covered the TLSA. The ITMG (◘ Figs. 4.21 and 4.22) provided protection against abrasion, thanks to layers of Teflon and Chromel-R.

Liquid Cooling Garment

The liquid cooling garment (LCG) was made of a spandex/nylon material and was worn next to the skin. The LCG (◘ Figs. 4.23 and 4.24) provided a steady flow of temperature-controlled water (which was supplied either from the LM or from the PLSS) through a network of polyvinylchloride (PVC) tubes that were integrated into the garment (◘ Table 4.12).

Lunar Boots

The lunar boot comprised two layers. The outer layer, except for the sole (which was made of Nomex), was made from Chromel-R and Teflon-impregnated beta cloth, whereas the inner layer comprised two Kapton layers and five Mylar layers separated by Dacron and Teflon-coated beta cloth [16].

Constant Wear Garment

The constant wear garment (CWG), which was made from cotton, was worn next to the skin during IVA CM operations. It served to provide astronauts with comfort and a means to absorb perspiration. A fly opening and a rear "port" allowed bodily functions to be performed.

Communications Carrier

The communications carrier, which featured microphones and earphones embedded in a skull cap, allowed astronauts to talk to one another and to mission control.

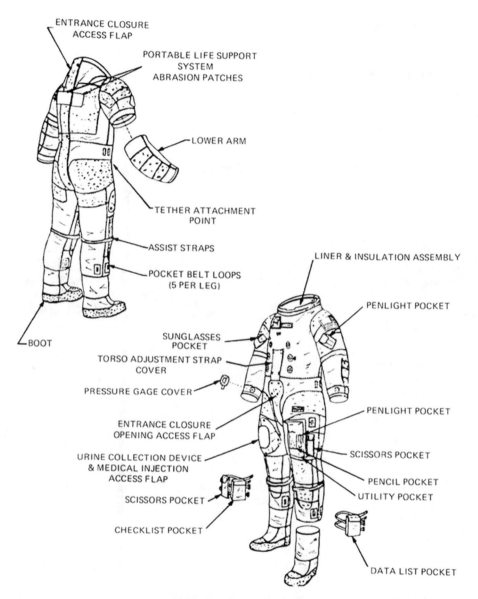

ENTRANCE CLOSURE
ACCESS FLAP

PORTABLE LIFE SUPPORT
SYSTEM
ABRASION PATCHES

LOWER ARM

TETHER ATTACHMENT
POINT

ASSIST STRAPS

POCKET BELT LOOPS
(5 PER LEG)

LINER & INSULATION ASSEMBLY

BOOT

SUNGLASSES
POCKET

TORSO ADJUSTMENT STRAP
COVER

PRESSURE GAGE COVER

ENTRANCE CLOSURE
OPENING ACCESS FLAP

URINE COLLECTION DEVICE
& MEDICAL INJECTION
ACCESS FLAP

SCISSORS POCKET

CHECKLIST POCKET

PENLIGHT POCKET

PENLIGHT POCKET

SCISSORS POCKET

PENCIL POCKET

UTILITY POCKET

DATA LIST POCKET

◘ **Fig. 4.21** The ITMG. Credit: NASA

Portable Life Support System

The PLSS comprised a backpack (◘ Fig. 4.25) that supplied the astronauts with oxygen for breathing, suit pressure control, removal of carbon dioxide, trace contaminants and odors, temperature control, and warnings of malfunctions (◘ Table 4.13). The PLSS also supplied oxygen to the PGA and water to the LCG. The subsystems included an oxygen ventilating circuit, a primary oxygen subsystem, a liquid transport loop, a feedwater loop, and a power system. The oxygen ventilating circuit

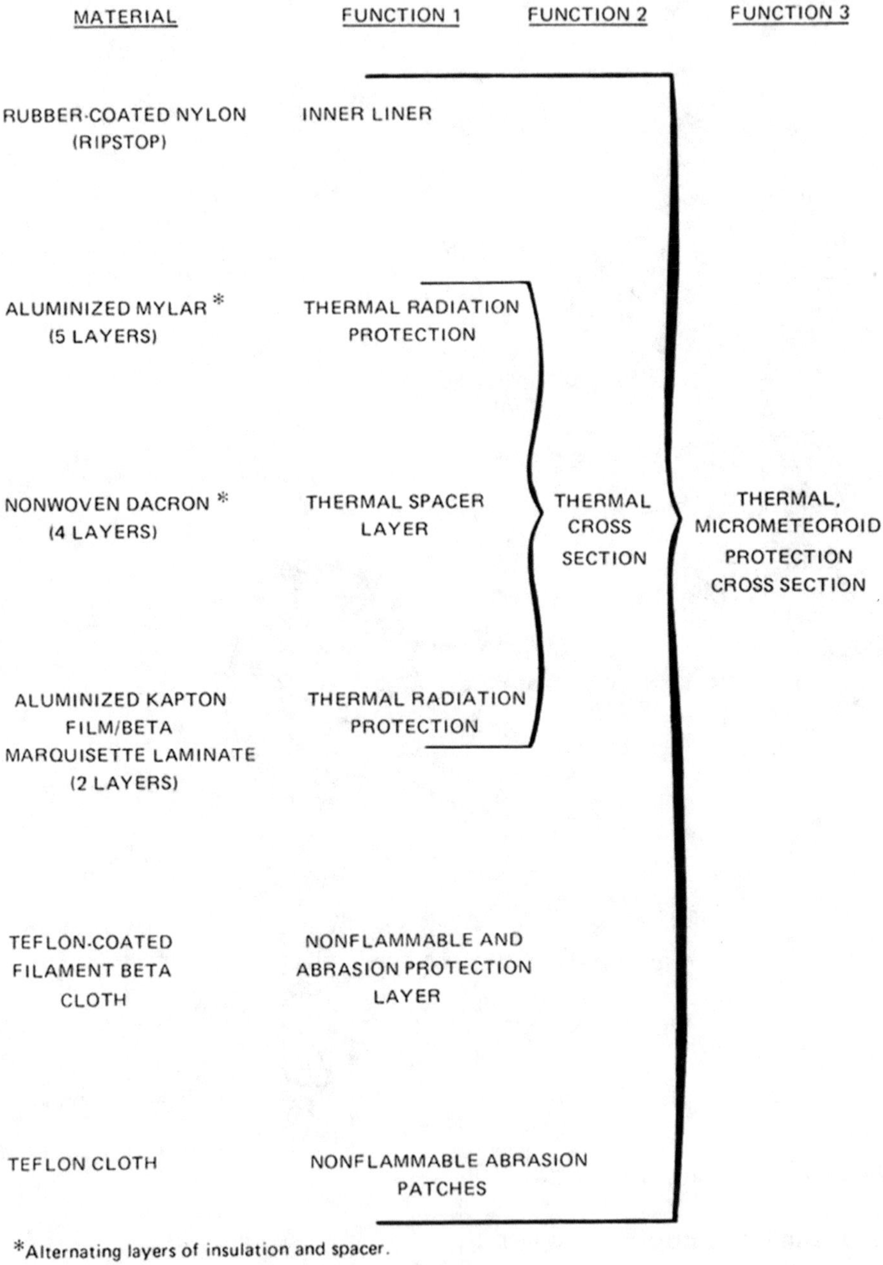

4

MATERIAL	FUNCTION 1	FUNCTION 2	FUNCTION 3
RUBBER-COATED NYLON (RIPSTOP)	INNER LINER		
ALUMINIZED MYLAR * (5 LAYERS)	THERMAL RADIATION PROTECTION		
NONWOVEN DACRON * (4 LAYERS)	THERMAL SPACER LAYER	THERMAL CROSS SECTION	THERMAL, MICROMETEOROID PROTECTION CROSS SECTION
ALUMINIZED KAPTON FILM/BETA MARQUISETTE LAMINATE (2 LAYERS)	THERMAL RADIATION PROTECTION		
TEFLON-COATED FILAMENT BETA CLOTH	NONFLAMMABLE AND ABRASION PROTECTION LAYER		
TEFLON CLOTH	NONFLAMMABLE ABRASION PATCHES		

*Alternating layers of insulation and spacer.

◘ **Fig. 4.22** Functions of the layers of the ITMG. Credit: NASA

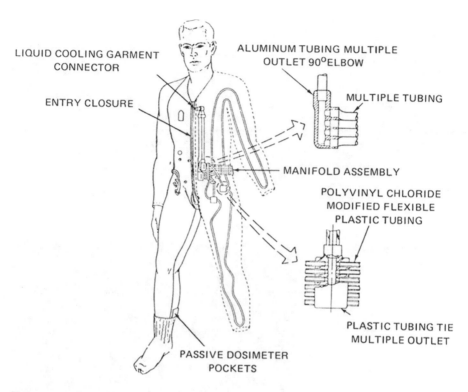

LIQUID COOLING GARMENT CONNECTOR

ENTRY CLOSURE

ALUMINUM TUBING MULTIPLE OUTLET 90°ELBOW

MULTIPLE TUBING

MANIFOLD ASSEMBLY

POLYVINYL CHLORIDE MODIFIED FLEXIBLE PLASTIC TUBING

PLASTIC TUBING TIE MULTIPLE OUTLET

PASSIVE DOSIMETER POCKETS

◘ **Fig. 4.23** The LCG. Credit: NASA

◘ **Fig. 4.24** LCG material. Credit: NASA

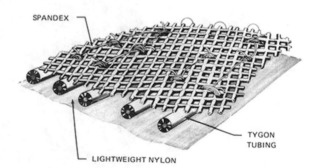

SPANDEX

TYGON TUBING

LIGHTWEIGHT NYLON

(◘ Fig. 4.26) provided contaminant control, regulated temperature, and ensured humidity stayed within nominal limits, while the primary oxygen subsystem supplied oxygen for pressurizing the suit and for breathing. The liquid transport loop provided temperature control; water from the LCG entered the loop via a water connector and was pumped through a sublimator so heat could be radiated away. The feedwater loop supplied any excess water to the sublimator for cooling. The final subsystem, the electrical power subsystem, powered the fan and pump motor assemblies, the communication system and instrumentation, which were located in the chest-mounted remote control unit shown in ◘ Fig. 4.27 [16].

4

◘ **Table 4.12**   Liquid cooling garment characteristics [16]

Weight (charged)	2.09 kg
Operating pressure	28 958 to 158 579 N/m^2
Structural pressure	217 185 ± 3447 N/m^2
Proof pressure	217185 ± 3447 N/m^2
Burst pressure	327 501 N/m^2
Leak rate, 131 000 N/m^2 (19 psig) at ≈ 280°K (45°F)	0.58 cm^3/hr

◘ **Fig. 4.25**   The Apollo PLSS. Credit: NASA

◘ **Table 4.13** PLSS specifications [16]

Design requirements	Specifications	
	Apollo 11–14	**Apollo 15–17**
Average metabolic load	6894 J/hr	6694 J/hr
Peak metabolic load	8368 J/hr	8368 J/hr
Maximum heat leak in	1046 J/hr	1255 J/hr
Maximum heat leak out	1046 J/hr	1464 J/hr
Maximum CO_2 partial pressure	15 mm Hg	15 mm Hg
Pressure garment assembly pressure	3.85 psia	3.85 psia
Ventilation flow	.1557 m³/min	.1557 m³/min
Duration	4 hr	7 hr
Oxygen charge pressure at $\approx$ 294°K (70°F)	1020 psia	1410 psia
Battery capacity	279 W-hr	431 W-hr
Emergency oxygen		
Duration (minimum)	30 min	30 min
Maximum flow	3.63 kg/hr	3.63 kg/hr
Pressure garment assembly pressure	3.7 psia	3.7 psia

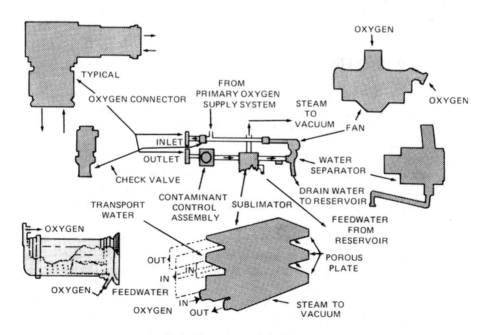

◘ **Fig. 4.26** The PLSS oxygen ventilating circuit. Credit: NASA

4

Of all the life support technology of the Apollo era, the EMU (◼ Fig. 4.28) was one of the most outstanding. Although the astronauts reported a few issues such as the lack of dexterity of the gloves and the tendency of the visor to scratch easily, the EMU never suffered even a minor failure. The EMU performed flawlessly during egress and ingress, it allowed astronauts the mobility and balance required to perform their lunar tasks, and it was so comfortable that some astronauts reported they almost forgot they were wearing a suit! One of the outstanding features of the suit was its endurance capability, as evidenced by the 7-hour 37-minute EVA of Apollo 17, which was the longest EVA on the Apollo Program. The performance of the Apollo LSS and the biomedical results of this program can be found in NASA publication SP-368 (Section VI: Systems for the LSS), Following the Apollo Progra, LSS technology continued to be developed during the Skylab and Apollo Soyuz Test Program missions, but the next evolution of LSS technology came about as a result of the Space Shuttle Program (◼ Fig. 4.29), which is the focus of the following section.

Space Shuttle

The Space Shuttle's life support system was divided into eight subsystems as follows:
- Air Revitalization System (ARS)
- Water Coolant Loop System (WCLS)
- Atmosphere Revitalization Pressure Control System (ARPCS)
- Active Thermal Control System (ATCS)
- Supply and Wastewater System (SWWWS)
- Waste Collection System (WCS)
- Airlock Support System
- Extravehicular Activity Mobility Units

Air Revitalization System

The ARS controlled relative humidity between 30 and 75 percent, maintained carbon dioxide at breathable concentrations, controlled temperature at comfortable levels, maintained adequate ventilation, and provided cooling to the avionics in the flight deck and mid-deck. The subsystem itself comprised a series of water coolant loops, cabin air loops, and a means of controlling pressure [18]. Very simply, cabin air was directed to a heat exchanger in the crew compartment, which was the site at which cabin air was cooled by the water coolant loops. The water coolant loops collected heat from a heat exchanger in the crew cabin and transferred this heat to the heat exchanger in the ATCS.

What happened to all the heat? This was rejected by the ATCS via Freon-21 coolant loops, cold plate networks, and heat sink systems. Another means of rejecting heat was via the radiator panels fitted inside the payload bay doors. Operationally, this system worked by rejecting most of the heat via the payload bay doors, but when the doors had to be closed prior to return, the flash evaporators took over until the Shuttle reached an altitude of 100,000 feet [18]. At this altitude, the flash evaporators couldn't operate due to atmospheric pressure, so ammonia boilers took over.

◘ Fig. 4.27 The PLSS instrumentation unit. This unit signaled the astronaut if any suit systems malfunctioned. Such malfunctions included low ventilation flow, low PGA pressure, high oxygen flow, and high carbon dioxide levels. Credit: NASA

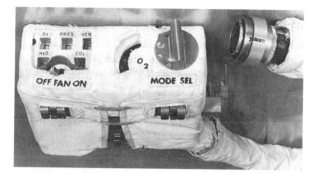

◘ Fig. 4.28 Apollo 17 astronaut Harrison Schmitt collects samples during the mission's first spacewalk at the Taurus-Littrow site. Credit: NASA

◘ Fig. 4.29 6 July 2006. Discovery approaches the International Space Station. The Leonardo Multipurpose Logistics Module can be seen in the payload bay. Credit: NASA

At the heart of the ARS were five air loops, which included one for the cabin, three for the avionics bays, and one for the inertial measurement units (IMUs). As air circulated through the cabin, it collected heat, odor, and carbon dioxide. The air was then drawn through the cabin loop through a filter and directed to lithium hydroxide canisters where carbon dioxide was removed. These canisters were removed every 12 hours (11 hours for a crew of seven). Cabin air was then directed through the cabin heat exchanger and cooled by the water coolant loops. The crew cabin had a volume of 2,300 cubic feet, which meant that with a flow rate of 330 cubic feet per minute, the crew cabin volume change occurred every 7 minutes [18].

Water Coolant Loop System

This subsystem collected heat at the cabin to air water coolant loop heat exchanger. The heat was transferred to the water coolant loops which transferred heat at the water and Freon-21 coolant loop interchanger. This series of thermal exchanges provided thermal regulation of the crew cabin, but the WCLS also provided thermal regulation for the avionics bays where heat generated by the operation of the avionics was transferred to a cold plate before being transferred to the WCLS [18].

Atmosphere Revitalization Pressure Control System

The ARPCS maintained cabin pressure at 14.7 psia (80 % nitrogen and 20 % oxygen) with an oxygen partial pressure maintained between 2.95 psia and 3.45 psia. Oxygen was supplied from two cryogenic storage systems located in the middle of the fuselage, while nitrogen was provided by two nitrogen tanks also located in the mid-fuselage. The nitrogen was also used to pressurize potable water located below the mid-deck floor. This water, which was produced by three fuel cell power plants, was used for crew consumption and also for the flash evaporator system [18].

The ARPCS was regulated by nitrogen and oxygen control and supply panels and partial pressure sensors that monitored positive and negative pressure using relief valves. On average, 1.76 pounds of oxygen was consumed per crewmember per day. The nitrogen tanks, which were constructed of Kevlar fiber, were pressurized to 3,300 psia. The concentration of nitrogen was controlled by supply valves, and the supply of the nitrogen to the cabin was regulated by reducing the pressure in stages. Control for this operation was performed via nitrogen controllers in the flight crew cabin. These controllers were normally set to "norm" or "auto" mode. In this latter mode, the system worked like this: when the partial pressure of oxygen sensor detected that oxygen was required, the nitrogen supply valve was simply closed automatically and oxygen flowed through its check valve to make up the required oxygen partial pressure [18]. To ensure the crew were warned of any off-nominal parameters, there was a red cabin atmosphere caution and warning light located on panel F7. This light was illuminated when the following parameters were exceeded:

- Cabin pressure was detected below 14 psia or exceeded 15.4 psia.
- Partial pressure of oxygen below 2.8 psia or above 3.6 psia was detected.
- Oxygen flow rate exceeded 5 pounds per hour.
- Nitrogen flow rate exceeded 5 pounds per hour.

In the case of a slow or rapid decompression, a klaxon would sound, and a master alarm light indicator would light up if pressure decreased faster than 0.05 psi per minute. To protect the crew from over-pressurization, relief valves were located in the crew cabin. To relieve over-pressurization, the crew simply switched the valves to "enable," and this permitted a relief of pressure of up to 150 pounds per hour [18]. Similarly, if a lower than normal pressure was detected, negative pressure relief valves were activated, which provided a flow of pressure into the cabin at up to 654 pounds per hour.

Active Thermal Control System

The ATCS removed heat from the ARS at the water coolant loop/Freon-21 coolant loop interchanger. This subsystem also provided heat for the PRSD cryogenic oxygen and oxygen supply line. The subsystem comprised two Freon-21 coolant loops (■ Fig. 4.30), which in turn comprised a cold plate network, a liquid heat exchanger, a heat sink system, a flash evaporator, and an ammonia boiler [18].

During orbital operations, once the payload doors had been opened, heat rejection was provided by (Freon) radiator panels (■ Fig. 4.31) in the payload bay doors. But during ground operations and from T+125 seconds (from lift-off

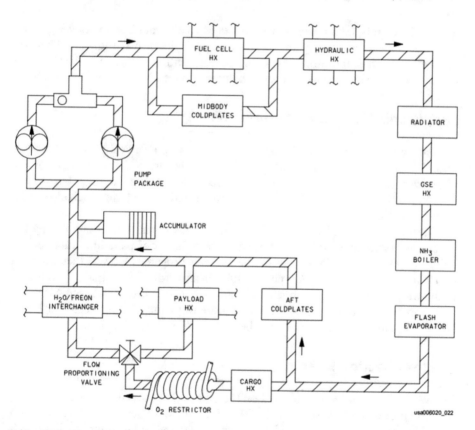

■ **Fig. 4.30** The ATCS Freon-21 coolant loop. Each loop comprised a pump package, which was basically two pumps and an accumulator. Credit: NASA

4

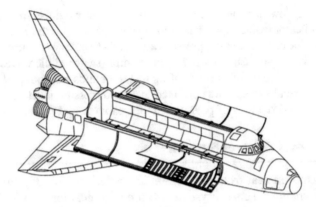

◻ **Fig. 4.31** Location of the Freon radiator panels. The radiators were located on the underside of the Orbiter's payload bay doors. On arriving on orbit, the doors were opened, and it was the radiator panels that were responsible for heat rejection. The radiator system, which comprised three panels (constructed of an aluminum honeycomb face sheet), was capable of heat rejection of up to 21,500 Btu per hour. Radiator flow control valve assemblies, which were located aft of the mod fuselage, were operated automatically or manually. Another element of the ATCS was ammonia boilers that cooled the Freon-21 coolant loops when the Orbiter was below 100,000 feet (during reentry). Credit: NASA

to T+125 seconds, heat rejection was achieved by thermal lag) to reaching orbit, heat rejection was enabled by the Freon-21 loops depicted in ◻ Fig. 4.30. Then, on reentry, the flash evaporator subsystem operated until an altitude of 100,000 feet was reached, at which point heat rejection was provided by ammonia boilers [18].

Supply and Wastewater System (SWWS)

This subsystem provided potable water for the crew and water for the flash evaporator. Water was generated by the fuel cells, which generated 25 pounds of potable water every hour. The potable water was stored in four tanks, each capable of storing 165 pounds (◻ Fig. 4.32). Wastewater was stored in one tank which also had a capacity of 165 pounds [18].

The water generated from the fuel cells was hydrogen-rich, which is why the water was first directed to hydrogen separators (a matrix of palladium tubes essentially), which removed 85 percent of the excess hydrogen which was then dumped overboard. After this, the water passed through a microbial filter that added iodine to the water. For consumption, chilled water was controlled to between 43°F and 55°F and for heated water, and the temperature was controlled between 155°F and 165°F. The ambient temperature was between 65°F and 95°F [18].

Waste Collection System

The WCS (◻ Fig. 4.33) was located on the mid-deck. It was used to collect, store, dry, and process crew waste. Fitted with a door, the door served the dual function as an ingress platform during prelaunch, when the crew had to enter the flight deck above the WCS.

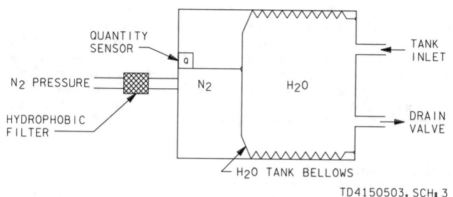

Fig. 4.32 Water tank schematic. The water tank was pressurized using nitrogen. A hydrophobic filter prevented any water that had leaked from getting into the water tank. Credit: NASA

Fig. 4.33 Performing ablutions in microgravity required some delicate maneuvering, as described by astronaut Mike Mullane: "NASA installed a camera at the bottom of the toilet simulator transport tube. A light inside the trainer provided illumination to a part of the body that normally didn't get a lot of sunshine. A monitor was placed directly in front of the trainer with a helpful crosshair marker to designate the exact center of the transport tube. In our training we would clamp ourselves to this toilet and wiggle around until we were looking at a perfect bull's-eye. When that was achieved, we would memorize the position of our thighs and buttocks in relation to the clamps and other seat landmarks." Credit: NASA

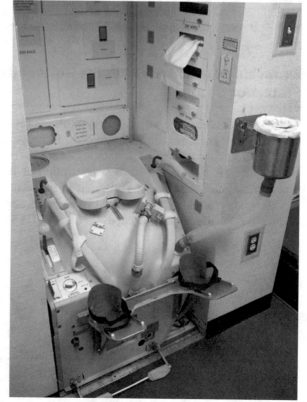

The WCS comprised a commode, a urinal, fan separators, odor filters, a vacuum vent disconnect, and controls. As you can see in ◘ Fig. 4.33, a hydrophobic bag liner was used for storing solid waste. In use, the commode was pressurized, and air flow was used to make sure waste moved in the right direction! The urinal was basically a funnel attached to a hose that used a fan separator to make sure liquid flowed where it was supposed to flow. As you can see, foot restraints were provided, and Velcro straps helped secure the feet of the crewmember using the WCS. Handholds and strategically positioned cameras were used for positioning [18].

4

Airlock Support

The airlock (◘ Figs. 4.34 and 4.35) was located on the mid-deck. It enabled astronauts to transfer from the mid-deck to the payload bay while wearing space suits without depressurizing the Orbiter. The airlock, which measured 63 inches in diameter and 83 inches in length and had a volume of 150 cubic feet, could accommodate two fully suited astronauts. It provided depressurization and repressurization, EVA equipment recharging, cooling for the liquid cooling ventilation garment, donning capabilities, and communication. To assist astronauts during pre- and post-spacewalk operations, the airlock featured handrails, foot restraints, and floodlights [18].

Extravehicular Activity Mobility Units and Crew Altitude Protection System

The EMUs provided the astronauts with all life support requirements, including supply of oxygen, removal of carbon dioxide, pressurized environment, temperature and humidity control, and micrometeoroid/orbital debris (MMOD) protection. ▶ Chapter 6 is dedicated to a description of the EMU and the prebreathe protocol. Before completing the overview of the Space Shuttle life support system, it is necessary to mention the Crew Altitude Protection System (CAPS) which later became the Advanced Crew Escape Suit (ACES) (◘ Fig. 4.36). This suit, manufac-

◘ **Fig. 4.34** The airlock. Credit: NASA

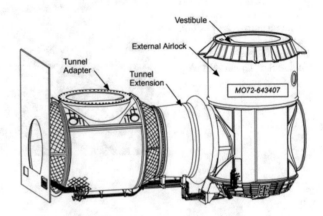

Fig. 4.35 A view of the airlock of Atlantis. Credit: NASA

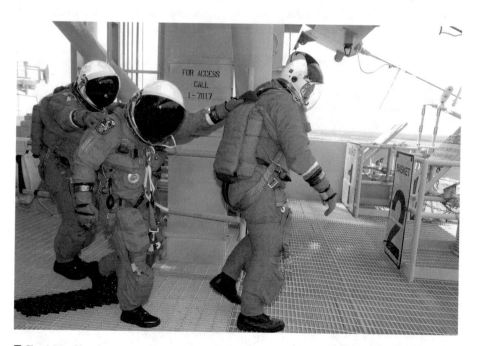

Fig. 4.36 Shuttle crewmembers practice an emergency egress in their ACES suits. Credit NASA

4

> ◼ **Table 4.14**   ACES features and capabilities
>
> Full pressure suit (launch and reentry), gloves, boots, helmet
>
> Provided 3.46 psi nominal operating pressure
>
> Provided protection in low altitude bailout and ground egress scenarios
>
> Was able to operate as open- or closed-loop demand breathing system
>
> Featured an emergency breathing system
>
> Liquid cooling system
>
> Headset communication system
>
> Search and rescue identification and emergency communication hardware
>
> High-altitude, automatic inflation parachutes
>
> Automatic inflation life preserver
>
> Survival drinking water packets

tured by the David Clark Company, was a pressure suit worn by astronauts during launch and reentry. The ACES was a full pressure suit that provided, in addition to the features listed in ◼ Table 4.14, liquid cooling, emergency breathing, and survival hardware in the event of an emergency scenario.

Key Terms
- Active Thermal Control System (ATCS)
- Advanced Crew Escape Suit (ACES)
- Air Revitalization System (ARS)
- Atmosphere Revitalization Pressure Control System (ARPCS)
- Commander (CDR)
- Command Module (CM)
- Command Module Pilot (CMP)
- Constant Wear Garment (CWG)
- Crew Altitude Protection System (CAPS)
- Environmental Control System (ECS)
- Extravehicular Mobility Unit (EMU)
- Extravehicular Activity (EVA)
- Ground Elapsed Time (GET)
- Inertial Measurement Unit (IMU)
- Integrated Thermal Micrometeoroid Garment (ITMG)
- Intravehicular Activity (IVA)
- Lithium Hydroxide (LiOH)
- Liquid Cooling Garment (LCG)

- Lunar Extravehicular Visor Assembly (LEVA)
- Lunar Module (LM)
- Lunar Module Pilot (LMP)
- Micrometeoroid Orbital Debris (MMOD)
- Orthostatic Intolerance (OI)
- Oxygen Supply and Cabin Pressure Control Section (OSCPCS)
- Portable Life Support System (PLSS)
- Pressure Garment Assembly (PGA)
- Pressure Helmet Assembly (PHA)
- Service Module (SM)
- Supply and Wastewater System (SWWS)
- Torso–Limb Suit Assembly (TLSA)
- Urine Collection and Receptacle Assembly (UCTA)
- Urine Transfer System (UTS)
- Water Coolant Loop System (WCLS)
- Water Management System (WMS)

❓ Review Questions

1. What is meant by the term *orthostatic intolerance*?
2. Explain the function of the Gemini spacecraft's primary oxygen system.
3. Explain the function of the Gemini spacecraft's temperature control subsystem.
4. What elements comprised the suit loop of the Gemini space suit?
5. List five functions of the Apollo ECS system.
6. What functions did the CM pressure suit circuit support?
7. What was the OSCPCS?
8. List five PGA characteristics.
9. What was the function of the TLSA?
10. What metabolic load did the Apollo PLSS support?
11. List the eight subsystems of the Shuttle LSS.
12. Explain how the Shuttle ARS accomplished its function.

References

1. Daues, K. (2006, Jan). A history of spacecraft environmental control and life support systems. https://ntrs.nasa.gov/search.jsp?R=20080031131.
2. McDonnell, A., & C Corp. (1962, Dec). Project mercury familiarization manual.
3. Grinter, K. (2000). *Project mercury overview.* http://www-pao.ksc.nasa.gov/history/mercury/mercury-overview.htm.
4. Swenson, L. S., Grimwood, J. M., & Alexander, C. C. (1966). *This New Ocean, A History of Project Mercury.* Houston: Scientific and Technical Information Division, Office of Technology Utilization, National Aeronautics and Space Administration. in *Time* Vol. LXXIII (1959).

4

5. Link, M. M. & United States. National Aeronautics and Space Administration. Scientific and Technical Information Division. (2006). *Space Medicine in Project Mercury*. Houston: Scientific and Technical Information Division, National Aeronautics and Space Administration.

6. Manned Spacecraft Center (U.S.). (1962). *Results of the Third U.S. Manned Orbital Space Flight, October 3, 1962*. Washington: National Aeronautics and Space Administration, Manned Spacecraft Center, Project Mercury; for sale by the Superintendent of Documents, U.S. Govt. Print. Off.

7. Voas, R. B. (1960). Project Mercury astronaut training program. *Symposium on psychophysiological aspects of space flight* http://docplayer.net/986308-Project-mercury-astronaut-training-program-robert-b-voas-nasa-space-task-group-langley-field-virginia-introduction.htmL.

8. Wheelwright, C. D. (1962). *Physiological Sensors for use in Project Mercury*. Houston: Manned Spacecraft Center, National Aeronautics and Space Administration.

9. Manned Spacecraft Center (U.S.). *Results of the First U.S. Manned Suborbital Space Flight, June 6, 1961* (p. 1961). Washington, DC: National Aeronautics and Space Administration.

10. Manned Spacecraft Center (U.S.). (1961). *Results of the Second U.S. Manned Suborbital Space Flight, July 21, 1961*. Washington: U.S. Govt. Print. Off.

11. Manned Spacecraft Center (U.S.). (1962). *Results of the First U.S. Manned Orbital Space Flight, February 20, 1962*. Washington: U.S. Govt. Print. Off.

12. Manned Spacecraft Center (U.S.). *Results of the Second U.S. Manned Orbital Space Flight, May 24, 1962* (p. 1962). Houston: National Aeronautics and Space Administration, Manned Spacecraft Center.

13. McDonnell Aircraft Corporation. (1966, April). *NASA project Gemini familiarization manual*.

14. Berry, & Catterson, A. D. (1967, Feb). Pre-Gemini medical predictions versus Gemini flight results. In *NASA Manned Spacecraft Center, Gemini Summary Conference, NASA SP-138* (pp. 201–215). Washington; Berry and others, Man's Response to Long-Duration Flight in the Gemini Spacecraft, in NASA Manned Spacecraft Center, Gemini Mid-Program Conference, NASA SP-121 (Washington, Feb. 1966), pp. 235–44.

15. Hughes, D. F., Owens, W. L., & Young, R. W. (1973). *Apollo command and service module environmental control system - mission performance and experience*. New York: ASME Paper No. 73-ENA29, American Society of Mechanical Engineers.

16. SP-368 Biomedical Results of Apollo.

17. Brady, J. C., Browne, D. M., Schneider, H. J., & Sheehan, J. F. (1973). *Apollo lunar module environmental control system - mission performance and experience*. New York: ASME Paper No. 73-ENA28, American Society of Mechanical Engineers.

18. SHUTTLE CREW OPERATIONS MANUAL. Rev. A, CPN-1 DATE: December 15, 2008.

Suggested Reading

Daues, K. (2006, Jan). *A history of spacecraft environmental control and life support systems*. https://ntrs.nasa.gov/search.jsp?R=20080031131.

Diamant, B. L., & Humphries, W. R. (1990). Past and present environmental control and life support systems on manned spacecraft. SAE Transactions. Vol. 99, Section 1. *Journal of Aerospace, Part 1*, 376–408.

Sivolella, D. The Space Shuttle: Technologies and Accomplishments (June 2017). Springer-Praxis.

International Space Station Life Support System

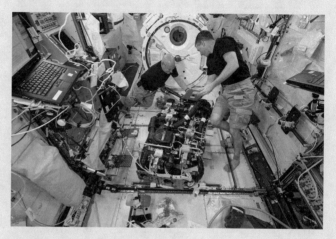

Credit: NASA

Contents

© Springer Nature Switzerland AG 2020
E. Seedhouse, *Life Support Systems for Humans in Space*,
https://doi.org/10.1007/978-3-030-52859-1_5

Learning Objectives

After reading this chapter, you should be able to:

- List each subsystem of the USOS and ROS LSS
- Describe the function of each subsystem of the USOS and ROS LSS
- Explain how the CDRA and CRA work
- Explain how the Sabatier reaction works
- Describe the function of the TCCS
- Explain how the CHX is used in the THC
- Describe the function of the NLSOV, VEDD, WPA, WLSOV, VEMRV, and VADD
- Explain the key features of the ROS Air Purification System and how each assembly achieves its function
- Explain what is meant by the *aggregation of marginal gains* in the context of LSS
- Explain how Thirsty Walls work

Introduction

This entire chapter is dedicated to the ISS life support system. Why? Well, for one thing, this spacecraft has been permanently crewed for more than two decades now, which means its life support system has been thoroughly and extensively tested. Over the years, life support engineers have gathered a wealth of data about the myriad subsystems and assemblies that comprise the orbiting outpost's life support system. Some subsystems, such as the oxygen generator assembly, have worked reliably, while others, such as the carbon dioxide removal assembly, have required extensive and intensive maintenance.

In addition to providing life support engineers with valuable data, the ISS has served as a vital test platform for the development of new technologies that will be required to close the system. Closure of the system is necessary if astronauts are ever to have a chance of venturing to Mars. Before reading this chapter, you should be advised that there are many, many acronyms, almost all of which you will be required to know since none of these are spelled out in any exam!

United States Orbital Segment Life Support System

The Environmental Control and Life Support System (ECLSS) for the International Space Station (ISS) includes some subsystems that are regenerative and some that are non-regenerative. Since a primary goal is to close the loop on the ECLSS, the aim has always been to increase the regenerative capabilities of these subsystems. During the 20 plus years the ISS has been orbiting Earth, there has been a gradual and steady increase in the closure of these subsystems [1], which, in the United States Orbital Segment (USOS)[1], are as follows:

1. **Atmosphere Control System**
   Assemblies: Manual Pressurization Equalization Valve
2. **Air Revitalization System**
   Assemblies: Carbon Dioxide Removal Assembly, Oxygen Generating Assembly, Trace Contaminant Control Assembly, Oxygen Recharge Compressor Assembly
3. **Temperature and Humidity Control System**
   Assemblies: Inter-module Ventilation, Common Cabin Air Assembly
4. **Fire Detection and Suppression System**
   Assemblies/Components: Portable Breathing Apparatus, Portable Fire Extinguisher
5. **Water Recovery Management System**
   Assemblies: Urine Processing Assembly
6. **Vacuum System**

Mass Balance

By increasing the regenerative capabilities of these subsystems, not only engineers are gradually closing the ECLSS loop, but they are also reducing the resources and consumables required. At this point, it is worth reminding ourselves of the mass balance (■ Fig. 5.1).

As we shall see in this chapter, many of the subsystems do a very good job at recycling water and generating oxygen, but there is still a 7 percent shortfall in overall closure, and this doesn't consider the food mass. Imagine a Mars-bound crew traveling to Mars with the life support system (LSS) currently in use on the ISS, and let's consider just the food mass penalty [1]. As you can see in ■ Fig. 5.1, each crewmember requires 800 grams of food per day, which equals to 3.2 kilograms per day for a crew of four. Multiply that by 1000 days, which is the average length of time for a Mars mission, and you have a mass penalty of 3200 kilograms. And we haven't even considered the shortfall in water and oxygen!

What is needed is a regenerative LSS, an element (water recovery) of which is depicted in ■ Fig. 5.2. The ISS LSS (■ Fig. 5.3) is partially regenerable.

Closing the loop is what NASA and the other space agencies are trying to do. As of 2020, the ISS ECLSS is classified as a partially closed LSS with a closure of 93 percent. According to NASA's website, there is no question of embarking on a trip to Mars until that number is closer to 98 or 99 percent. But it's not just a case of closing the loop. Engineers must also look at ways of enhancing LSS capability and extending the life of current subsystems. After all, there is no point in develop-

---

[1]The USOS comprises all module forward of the Pressurized Mating Adapter-1 (PMA-1): Unity, Permanent Multipurpose Module, Quest, Tranquility, the Cupola, Destiny, Harmony, Columbus, and the Japanese Experiment Module (JEM).

Life Requirements on Earth and in Space				
Item	On Earth		In Space	
	kg per person per day[1]	gallons per person per day	kg per person per day[2]	gallons per person per day
Oxygen	0.84		0.84	
Drinking Water	10	2.64	1.62	0.43
Dried Food	1.77		1.77	
Water for Food	4	1.06	0.80	0.21

Fig. 5.1 Daily mass balance for the average male astronaut. Note: these values obviously change depending on the mass of the astronaut. Credit: NASA

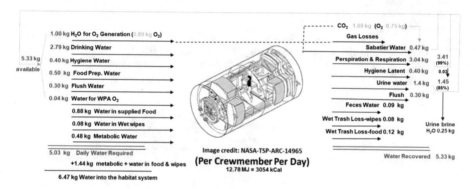

Fig. 5.2 Water mass balance per astronaut per day for a spacecraft with a regenerative life support system. Credit: NASA

ing an oxygen generating assembly that generates all the oxygen needs of the crew if that assembly breaks down every 2 weeks! So, let's begin with an overview of the ECLSS arrangement on ISS.

ISS ECLSS Overview

The ISS International Partners provide various ECLSS components within their modules, but not every module is fitted with a complete suite of LSS subsystems. For example, the Italian Space Agency's Permanent Multipurpose Module (PMM) provides only ventilation for the module and relies on adjacent modules for cooling and temperature control. A similar approach is implemented in several of the Russian Orbital Segment (ROS) modules for ACS, ARS, and WRM functions (**Fig. 5.4**).

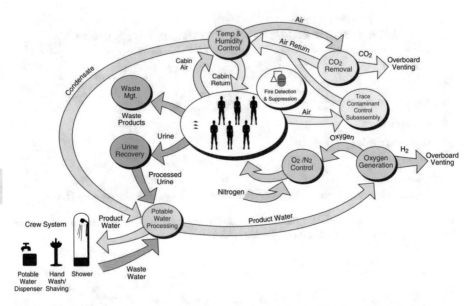

◘ Fig. 5.3 Schematic of a regenerative life support system. Credit: NASA

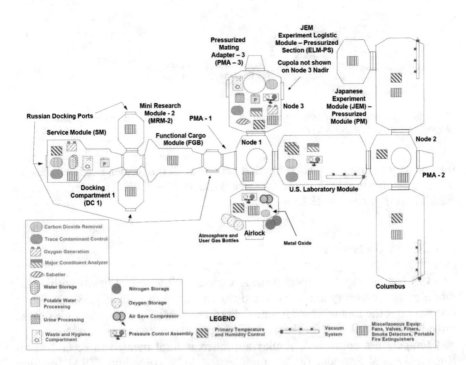

◘ Fig. 5.4 Schematic of various elements of the ISS ECLSS. Credit: ESA/NASA

Atmosphere Control System

One of the primary functions of the ACS is to maintain pressure. To do this requires an assortment of hardware such as sensors, distribution lines, and equalization valves. For example, the Node 1 ACS hardware includes pressure sensors that monitor the atmosphere pressure and oxygen and nitrogen distribution lines that ensure sufficient oxygen and nitrogen are pumped into not only this module but other modules such as Node 3, the US Lab, and the Quest module, which is where the airlock is located [2, 3, 4]. Another piece of hardware is the manual pressure equalization valve (MPEV). The MPEVs, which are attached to each of the hatches in the Node 1 module, are designed for bidirectional flow and can be operated manually from either side of the hatches [5, 6].

Air Revitalization System

The Air Revitalization System (ARS) subsystem is responsible for removing carbon dioxide, generating oxygen, and trace contaminant control. It achieves these functions via the carbon dioxide removal assembly (CDRA), the oxygen generation assembly (OGA), and the trace contaminant control system (TCCS). The CDRA was designed with the carbon dioxide reduction assembly (CRA) in mind. The CDRA collects carbon dioxide and directs it to the CRA, while the OGA electrolyzes water into oxygen and hydrogen. The oxygen is used by the crew to breathe while the hydrogen is directed to the CRA, which reacts with the hydrogen and carbon dioxide to generate methane and water [7]. Water is then directed to the OGA, which means this particular subsystem increases closure of the ECLSS (◘ Fig. 5.5).

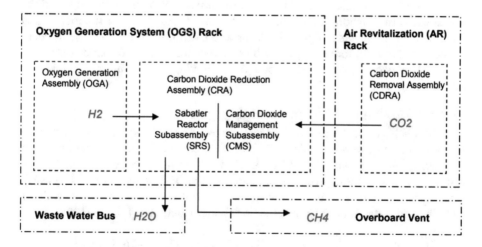

◘ **Fig. 5.5** Overview of ARS subsystems. Credit: NASA

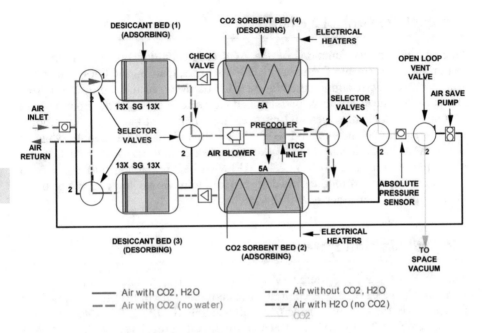

◘ Fig. 5.6 Four-bed molecular sieve schematic. Credit: NASA

Carbon Dioxide Removal Assembly

The CDRA removes carbon dioxide using a pressure and temperature swing adsorption cycle. Basically, the configuration of the CDRA includes carbon dioxide removal beds that are filled with a molecular sieve packed with heater plates (◘ Fig. 5.6). As more and more carbon dioxide is passed through the molecular sieve, the mass fraction of carbon dioxide gradually increases, and the bed adsorbs the carbon dioxide [8]. This process occurs because the material in the molecular sieve has a high affinity for water vapor. And because the molecular sieve material preferentially adsorbs and displaces carbon dioxide, it is necessary to use two desiccant beds, one for removing water vapor and one for replenishing water.

Carbon Dioxide Reduction Assembly

The CRA helps close the ECLSS ISS loop by using waste products that would otherwise be vented overboard. The CRA, which became fully operational on ISS in June 2011, interfaces with the OGA and CDRA. Carbon dioxide from the CDRA is stored in tanks until hydrogen is generated by the OGA via water electrolysis [9]. Once carbon dioxide is available, the CRA is activated, and the assembly produces methane and water, thanks to the reliable Sabatier reaction as follows [10]:

$$\text{Sabatier Reaction}: CO_2 + 4H_2 \Leftrightarrow CH_4 + 2H_2O \quad \Delta H^\circ_{rxn} = -165 \, kJ/mol \qquad (1)$$

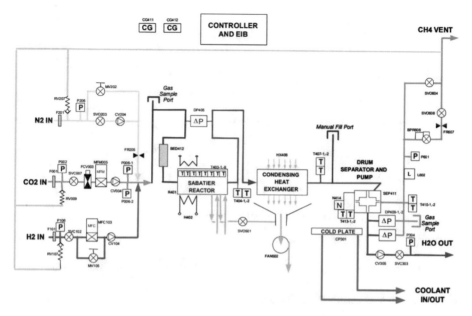

◻ Fig. 5.7 How the Sabatier reaction works in the CRA. Credit: NASA

For those who have a hard time figuring out formulae and are wondering how the Sabatier reaction works, here goes. First, water is condensed out of the stream generated by the OGA. It is then separated and purified in the water processing assembly (which we will discuss when we describe the Water Recovery Management System later). Once purified, water is recycled back to the OGA, where it used to generate oxygen for the crew to breathe. Methane, which is saturated with water vapor at this point, is at a dew point temperature that is very close to the ISS cooling loop used to cool the station's condensing heat exchanger, as depicted in ◻ Fig. 5.7. This methane is simply vented to space.

Oxygen Generating Assembly

Another key feature of the ACS is the oxygen generating assembly (OGA), the function of which is to convert potable water from the Water Recovery Management System into oxygen and hydrogen. The oxygen is used to make up the breathable atmosphere in the modules, and the hydrogen either is used to generate more water or is simply vented to space. The OGA (◻ Fig. 5.8), which was activated in July 2007, has generally proven very effective at generating oxygen but has shown to be rather maintenance-intensive; filters have to be replaced regularly, the system has to be calibrated, sensors need to have maintenance checks, loop filters have to be replaced, pumps have to be checked, seals must be inspected, etc. You get the idea.

In fact, the OGA has been plagued by clusters of problems (◻ Figs. 5.9, 5.10, and 5.11) caused by design oversight, which in turn was caused by lack of sufficient

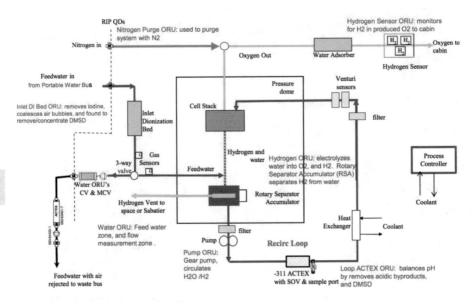

■ **Fig. 5.8**   OGA schematic. Credit: Tanada

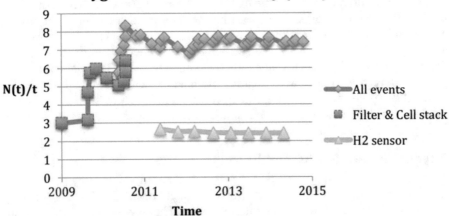

■ **Fig. 5.9**   The cumulative failure and maintenance rate for the combined filter and cell stack group and for the H2 sensor group in the OGA. Credit: NASA / Tanada

ground testing. In short, the cause of all the failures observed during the time the OGA has been installed on the ISS can only be solved by a major design modification. In other words, you don't want to be taking this along with you to Mars! That is because while the OGA does a great job at generating/recycling oxygen, this generating/recycling benefit is overshadowed by the sheer number of maintenance and repair cycles.

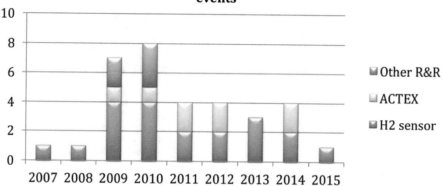

○ **Fig. 5.10** The number of OGA repair and replace events. Credit: NASA / Tanada

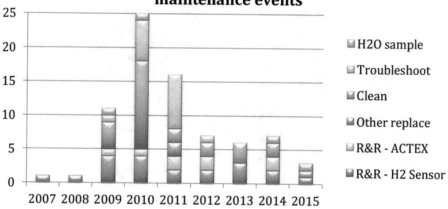

○ **Fig. 5.11** The number of OGA maintenance events each year. Credit: NASA (Tanada)

Oxygen Recharge Compressor Assembly

The oxygen recharge compressor assembly (ORCA) is a means by which the ISS stores oxygen on board. When depleted, these tanks, which are pressurized at 2400 psi, are either replaced or repressurized.

Trace Contaminant Control Assembly

The trace contaminant control assembly (TCCS) achieves its function of removing trace chemical contaminants using the processes of physical adsorption, thermal catalytic oxidation, and chemical adsorption. Located in the ARS rack inside the Destiny module, the TCCS (○ Figs. 5.12 and 5.13) features a charcoal bed assembly (CBA), a thermal catalytic oxidizer assembly (COA), and a post-sorbent bed

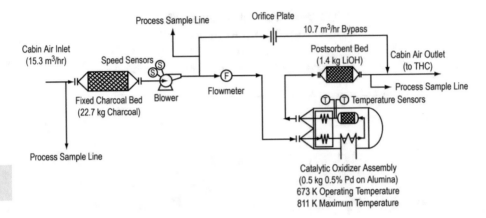

■ **Fig. 5.12**   Simplified TCCS process. Credit: NASA

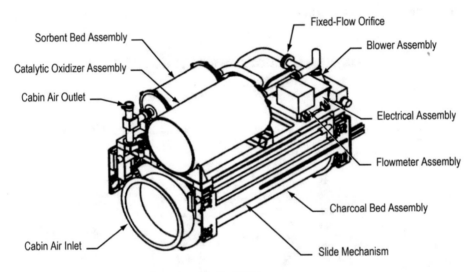

■ **Fig. 5.13**   TCCS flight configuration. Credit: NASA

assembly (SBA). It is the CBA, COA, and SBA that are the key components that remove contaminants. The process begins as cabin air is directed into the TCCS at a rate of 15.29 cubic meters per hour [11]. The air is then directed through the CBA, which is an expendable fixed bed that contains about 22 kilograms of Barnebey Sutcliffe Corporation Type 3032 charcoal. As the air passes through the CBA, the charcoal removes many of the volatile organic compounds (VOCs), but it cannot remove all the methane or carbon monoxide. But not to worry, because the COA can do this job, thanks to its integrated heat exchanger and its bed of catalyst pellets. This bed comprises 500 grams of alumina pellets. The COA, which receives 4.59 cubic meters of air per hour, is maintained at 400°C, a temperature sufficient to ensure the process of VOC oxidation is achieved. Finally, after exiting the COA, air flows through a fixed bed of granular lithium hydroxide (LiOH), which breaks down the remaining VOCs, and ultimately, the SBA provides post-treatment of the resultant air.

Advances in ARS Technology: The Advanced Closed-Loop System

The advanced closed-loop system (ACLS) is a regenerative LSS subsystem. Its regenerative processes include removing carbon dioxide, generating oxygen, and reprocessing carbon dioxide [7]. It achieves these functions via:

1. The Carbon Dioxide Concentration Subsystem (CCA), which controls carbon dioxide levels inside the module
2. The Carbon Dioxide Reprocessing Subsystem (CRA), which uses the Sabatier reaction to react hydrogen and carbon dioxide over a catalyst to generate water and methane.
3. The Oxygen Generation Subsystem (OGA), which splits water into oxygen and hydrogen

The ACLS (◘ Fig. 5.14), which is currently operating as a technology test and demonstrator, represents a significant step forward in regenerative LSS. It can process 40 percent of the water required for producing oxygen from the carbon dioxide exhaled by the crew.

Temperature and Humidity Control System

The temperature and humidity control (THC) system's primary function of regulating temperature and humidity is achieved through the use of ventilation fans, a low-temperature cooled hydrophilic-coated heat exchanger (CHX), and a rotary liquid separator that removes condensate from the stream of air. In terms of performance and reliability, the system has performed well over time, although there is a tendency for sloughing of the CHC coating material. The THC's secondary function is cleaning the air of particulates and microbes, and it achieves this function with the use of HEPA filters. This is a simple solution, but as you can imagine, the filters are expendable and must be replaced regularly.

Ventilation capability is achieved by two processes:

1. Exchanging air between the modules using inter-module ventilation (IMV) [12, 13, 14]. For example, the arrangement of ventilation elements and ducting ensures air is exchanged between the Russian Orbital Segment (ROS) and the United States Orbital Segment (USOS).
2. Circulating air within a module. One way to understand how this works is to focus on one particular module. We'll take a look at Harmony, which is where the crew quarters (CQ) are located. To provide each crewmember with comfortable air temperature and humidity, air is circulated through each CQ by two fans. Inside the CQ, air adsorbs heat from the metabolic heat generated by the astronaut. The air is then directed through a CQ exhaust duct and toward the common carrier air assembly (CCAA).

5

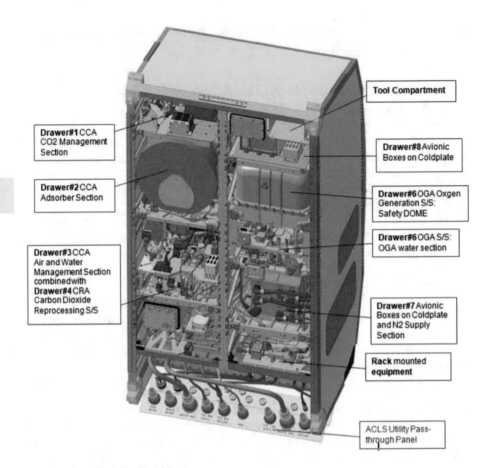

Drawer#1 CCA
CO2 Management
Section

Tool Compartment

Drawer#8 Avionic
Boxes on Coldplate

Drawer#2 CCA
Adsorber Section

Drawer#6 OGA Oxgen
Generation S/S:
Safety DOME

Drawer#6 OGA S/S:
OGA water section

Drawer#3 CCA
Air and Water
Management Section
combined with
Drawer#4 CRA
Carbon Dioxide
Reprocessing S/S

Drawer#7 Avionic
Boxes on Coldplate
and N2 Supply
Section

Rack mounted
equipment

ACLS Utility Pass-
through Panel

◘ Fig. 5.14   ACLS. Credit: ESA

Fire Detection and Suppression System

The fire detection and suppression system (FDS) does exactly that: detects fires and suppresses them. It achieves these functions using a photoelectric fire detector, a portable breathing apparatus (PBA), and a portable fire extinguisher (PFE). The PFE is a standard mask that covers the face and is attached to a 6-meter hose that is plugged into an air socket. The PFEs come in two varieties: one using carbon dioxide, weighing 2.7 kilograms, and the other using water mist – the fire water mist PFE (FWM PFE) (◘ Fig. 5.15) [15, 16, 17].

Water Recovery and Management System

The water recovery and management (WRM) system must receive wastewater such as crew urine and condensate, then process that wastewater to drinking water standards through the water recovery system (WRS), and distribute that water back to the astronauts [18, 19]. A schematic showing how these functions are achieved is depicted in ◘ Fig. 5.16.

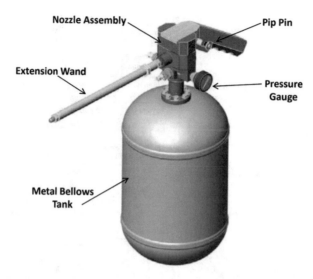

◘ Fig. 5.15 The FWM PFE. This PFE comprises a nozzle assembly and a tank assembly. The nozzle comprises the handle, cartridge valves, nozzle tip, and venturi, while the tank, which is titanium, contains 6 pounds of water and 1.2 pounds of pressurized nitrogen. To operate, the astronaut simply removes the pip pin and squeezes the handle. This opens the two cartridge valves that permit nitrogen to flow and water to mix in the venturi before being discharged through the nozzle [15, 16, 17]. Fortunately, as of June 2020, these PFEs have not been used. Of all the LSS subsystems, the PFEs are the least maintenance-intensive, since they have 15-year shelf lives. Credit: NASA

Here's how the whole process works. First, the wastewater bus receives humidity condensate from the CCAA. Remember, it is the function of the CCAA, which is a THC assembly, to condense water vapor and other condensable contaminants to this bus by first directing the condensate through a water separator.

In addition to utilizing the CCAA, the WRM makes use of the carbon dioxide reduction system (CDRS). Remember, the CDRS produces water from the Sabatier reaction to produce water from carbon dioxide from the CDRA (this assembly also produces hydrogen from electrolysis in the OGA). All this water is routed to the water processing assembly (WPA) waste tank. This is a different tank than the waste and hygiene compartment (WHC), which collects urine. It is in the WHC that urine is treated with chemicals. Once pretreated, it is routed to the UPA (◘ Figs. 5.17 and 5.18) where vapor compression distillation (VCD) technology recovers water from urine by evaporating it at low pressure in the distillation assembly. The efficiency with which it does this is about 74 percent. Incidentally, the amount of time it takes for urine to become water again for drinking is about 8 days on station. And yes, the water is very pure. In fact, it far exceeds the purification standards imposed by most local authorities on Earth. Now you may be wondering why the UPA only functions at 74 percent. Well, the manufacturer confidently predicted it would operate at more than 90 percent and when, after a few months of operating on ISS, that number was below what was predicted, an investigation was launched. Seems the company that designed the UPA had not considered the effect of bone loss when designing their system. You see, when astronauts are in space, calcium leaches out of their bones and that calcium is filtered through the

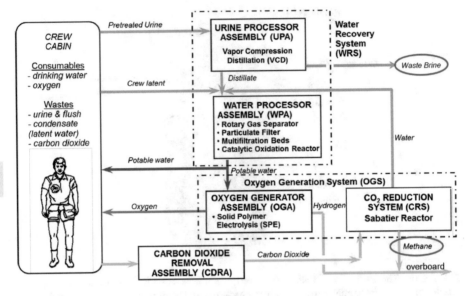

Fig. 5.16   Function of the WRM. Credit: NASA

Fig. 5.17   The UPA.
Credit: NASA

kidney in the urine. And when you try to pass calcium-rich urine through a filter, the filter clogs!! Hence the reduced efficiency and a valuable lesson in UPA design!

Vacuum System

The vacuum system is the subsystem that is least mentioned when talking about LSS. It is often used to conduct experiments. This subsystem is located in Columbus, the JEM, and Destiny. The function of this subsystem is to provide vacuum, venting, and distribution of nitrogen. It achieves these functions, thanks to myriad valves and shut-off valves. The nitrogen line shut-off valves (NLSOV) are used to provide nitrogen to the International Standard Payload Racks (ISPR) and to the water pump assembly (WPA) for repressurization. The Venting Dump Devices

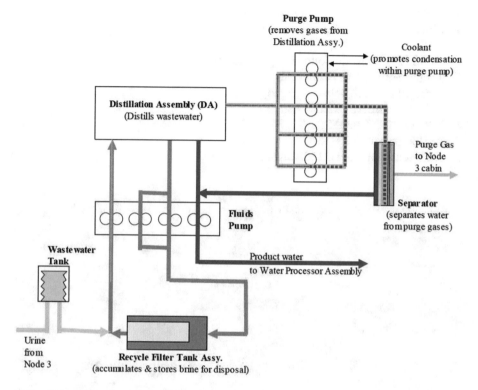

⬛ Fig. 5.18 UPA operation. Credit: NASA

(VEDD), on the other hand, are used to dump waste gas into space. This action can only be achieved by first opening the waste line shut-off valves (WLSOV). Once the WLSOV and VEDD have been closed, vent lines are repressurized using cabin air by opening the venting manual return valves (VEMRV). A number of racks inside the JEM, Columbus, and Destiny are equipped with vacuum lines to enable vacuum conditions for payloads. To enable this function, Vacuum Dump Devices (VADD) are opened to dump air inside the vacuum line to space. The vacuum line is then repressurized using the same procedure as the venting line.

That completes the overview of the USOS LSS. From a mission control perspective, the LSS, the ETHOS acronym is applied. This stands for Environmental and Thermal Operating System, and for a more detailed insight into how the system is operated, the Console Handbook is provided in Appendix IV.

Russian Orbital Segment Life Support System

The ROS LSS provides the same functions as the USOS LSS, although the organization is a little different. There are five subsystems as follows:

1. Atmosphere control system
2. Water supply system
3. Food supply facilities

4. Sanitation and hygiene compartment
5. Fire detection and suppression

The ROS LSS fulfils the following parameters (□ Table 5.1):
1. Total pressure: 660–860 mmHg
2. PPO$_2$ inhaled air: 150–200 mmHg
3. PPCO$_2$: 6 mmHg
4. PPH$_2$O: 10 ± 5 mmHg
5. Air temperature: 20–25°C
6. Air circulation rate: 0.1–0.4 m/sec

5

□ **Table 5.1**   Location of ROS ECLSS Units

Name	Zvezda	Zarya
Atmosphere control system		
Elektron unit	Panels 429, 430	
Solid fuel oxygen generator	Panel 429	
Vozdukh carbon dioxide removal system	Panels 420–424	
Carbon dioxide absorber canisters	Panels 432, 436	
Harmful impurities filter		Panel 411
Gas analysis system	Panel 439	Panel 405
Water supply system		
Rodnik system valves panel	Panels 233, 234	
Atmospheric condensate water regeneration system (БРП)	Panel 431	
Food supply facilities		
Food rations	Panels 238, 239	
Electric food warmer	In dining table	
On-board refrigerator	Panel 230	
Sanitation and hygiene equipment	Toilet compartment	
Fire detection and suppression system		
"Signal" system control panel	Panel 329	
Portable fire extinguisher	Hatch PO-ПРК Hatch PO-ПХО	Panels 229, 404 and in adapter
Rebreather-type gas mask	Panels 416, 216	Panels 230, 404 and adapter

Atmosphere Control System

The atmosphere control system is designed to supply oxygen to the crew and remove carbon dioxide and trace contaminants. This system must also monitor gas partial pressures and signal the crew if partial pressures exceed thresholds or fall below preset limits. Additionally, this system must equalize pressure between modules and pressurize the airlocks. To achieve these functions, this subsystem is organized into secondary subsystems as follows:

1. Oxygen supply system
2. Air purification system
3. Gas analysis system
4. Habitable compartment pressure integrity monitoring system
5. Interface pressure integrity monitoring system
6. Temperature and humidity control

Oxygen Supply System

This system comprises the Elektron unit (⬛ Fig. 5.19) and two solid fuel oxygen generators. The Elektron, which generates oxygen by means of the electrochemical decomposition of water, is the primary source of oxygen. The unit comprises a liquid

⬛ **Fig. 5.19** Andrei Borisenko (L), Alexander Samokutyaev (C), and Sergey Volkov (R) display the Elektron oxygen generator in Zvezda. Credit: NASA

loop containing 30 percent potassium hydroxide, gas lines, and electromagnetic valves. The liquid loop features an electrolysis unit, heat exchangers, a pump, and a water storage tank. The pump circulates the electrolyte via the electrolysis unit, and oxygen (and hydrogen) is generated by the electrolysis of water in the potassium hydroxide solution. The oxygen is released directly into the module, while the hydrogen is vented into space. The unit is quite effective, decomposing 1 kilogram of water per hour which yields 25 liters of oxygen per hour, which is sufficient to support one astronaut or cosmonaut for 1 day. Controlling the Elektron is a computer which monitors valve status, oxygen pressure, hydrogen concentration, and oxygen concentration. If an exceedance is registered, a signal is sent to the Integrated Control Panel, and if a threshold is exceeded, then an automatic shutdown is triggered.

The two solid fuel generators comprise a replaceable cartridge with igniter, a striker mechanism, a filter, and a fan. The generators, which are activated if the partial pressure of oxygen falls below 160 mmHg, are designed for thermal decomposition of an oxygen compound that is packed into a cartridge. Each cartridge generates 600 liters of oxygen, and the contents take up to 20 minutes to decompose.

Air Purification System

This system removes carbon dioxide and trace contaminants from the atmosphere. The system includes the following assemblies:
1. Vozdukh carbon dioxide removal system
2. Carbon dioxide absorbent canisters
3. Trace contaminant control unit
4. Harmful contaminants control filter

Vozdukh Carbon Dioxide Removal System

The Vozdukh comprises three parts: a preliminary purification unit, a heat exchanger, and an atmosphere purification unit. This system achieves its function of removing carbon dioxide by removing the gas by molecular sieve beds, which comprise zeolite, a porous, adsorbent material. The sieve beds operate via capillary action, and the efficiency with which carbon dioxide is removed is dependent on the air flow rate, the concentration of carbon dioxide in the atmosphere, and the length of the adsorption/desorption cycle. Once the zeolite is saturated, the material is regenerated by simply exposing the bed to vacuum. The system can operate in automatic and manual modes, and as with all LSS systems, preset thresholds ensure the partial pressure of carbon dioxide is maintained within nominal parameters.

Carbon Dioxide Absorbent Canisters

The carbon dioxide absorbent canisters are the backup to the Vozdukh. As with most carbon dioxide LSS assemblies the canisters remove carbon dioxide using LiOH.

Trace Contaminants Control Unit

This system, which comprises two regenerable activated charcoal cartridges, a catalytic oxidizer, a filter, a fan, and valves, can be operated using either of two very simple modes: one cartridge at a time or both cartridges simultaneously. In use, one fan draws air through the filter and through the cartridges, which adsorb the carbon dioxide and contaminants.

Harmful Impurities Filter

This system, which comprises a replaceable cartridge (containing a chemical sorbent and activated charcoal) and a catalyst (that oxidizes carbon monoxide to carbon dioxide), absorbs gases such as ammonia, hydrogen sulfide, and hydrocarbons.

Gas Analysis System

The gas analysis system, of which there are two (one for Zvezda and one for Zarya), ensures round-the-clock monitoring of partial pressures of oxygen, carbon dioxide, and hydrogen. The system operates within the following ranges:
1. Partial pressure of oxygen: 0–350 mmHg
2. An alarm sounds if PPO_2 falls below 120 mmHg
3. Partial pressure of carbon dioxide: 0–25 mmHg
4. An alarm sounds if $PPCO_2$ rises above 20 mmHg
5. Partial pressure of water vapor: 0–30 mmHg
6. Hydrogen content: 0–2.5 %

Habitable Compartment Pressure Integrity Monitoring System

This system comprises various pressure monitoring devices/sensors linked to the computer system, which is responsible for processing the data from the sensors and ensuring parameters are within preset thresholds. If a low pressure drop rate is detected, a warning signal is sent to the caution and warning panel, and a yellow indicator light starts flashing. If a high pressure drop rate is detected, a red indicator light starts flashing, and an audio alarm sounds.

Interface Pressure Integrity Monitors

This system comprises a series of pressure equalization valves, tunnel pressure monitoring valves, and a vacuum manometer. The pressure equalization valves, which are installed on pressurized bulkheads, are used to pressurize the docking nodes. Information about the status of these valves is indicated on the Integrated Control Panel. The tunnel pressure monitoring valve, which is controlled using a two-position handle, is connected to the vacuum manometer to the chamber of the node that is to be tested. The vacuum manometer is an aneroid device that measures total pressure in the ROS.

Temperature and Humidity Control

Temperature and humidity control is achieved using liquid-air heat exchangers. In this system, air is cooled and heated automatically by two thermal control system loops. Since convection does not occur naturally on the ISS, this system uses ventilation equipment to stir the air in the ROS.

Water Supply System

5

The ROS water supply system stores and supplies potable water, regenerates condensate, and provides water for crew hygiene. These functions are achieved by water delivery through cargo vehicles (such as the Progress) and water regeneration. On arrival at the ISS, water on board the Progress is pumped into the Rodnik system located in Zvezda. It is the Rodnik system, which simply comprises two water tanks, that provides the crew with potable water. Each tank can hold 210 liters of water and the shelf life of the water is 365 days.

Atmospheric Condensate Water Regeneration System

This system generates potable water. Moisture that forms inside the ROS collects as it condenses on the cold surfaces of the temperature and humidity control system's heat exchangers. The temperature and humidity control system directs the condensate to the water regeneration system, which processes it into potable water. Part of the processing includes purifying the water by passing it through columns filled with resins and charcoal. Before the water is pronounced potable, it must pass through a quality sensor that checks for contaminants. The system, which operates around the clock, produces 1.2 to 1.3 liters per crewmember per day. This water is received by the cosmonauts via dispenser valves of a distribution and heating unit.

Food Supply Facilities

These facilities, which are located in Zvezda, help cosmonauts store and prepare food in addition to providing a means of collecting and storing food waste. Each day, cosmonauts are provided breakfast, snack, lunch, and dinner with a combined calorific value of 3000 kcal. This food, which has a shelf life of 240 days, is stored in containers at ambient temperature or in the refrigerator. Freeze-dried foods must be reconstituted with hot (72–83°C) or cold (10–45°C) water from the condensate water regeneration system, whereas food in cans (and bread) may be warmed to 65°C in the electric food warmer, of which there are two in Zvezda. So, say you fancy some warm bread. What do you do? Well, first you would remove the bread from the outer plastic bag, insert the bread into the food warmer, dial the temperature up to 65°C, and wait for around 30 minutes. Once the heating cycle is complete, you would simply turn the heater off and remove the bread.

Refrigerator

This device stores food at temperatures between -5 and +10°C. The fridge is maintained once every 2 months, a process that requires all food to be removed and turning the fridge into drying mode for 6 hours.

Sanitation and Hygiene Equipment

This system collects and preserves metabolic waste and collects water used for washing hands. The system, which comprises a commode, urinal, and hand-wash station, is located in Zvezda. It comprises the following elements:
1. Solid waste container
2. Urine collector
3. Chemical pretreat pump
4. Conserving agent (sulfuric acid) tank
5. Flush water pump
6. Dynamic gas–liquid separator
7. Urine container
8. Urine/water storage container
9. Odor absorption filter
10. Fan
11. Hand-wash control panel
12. Hand-operated pump
13. Water storage container

How does it work? First, you need to switch the fan on to make sure everything moves in the right direction! Then, you complete your business, and the air flow created by the fan draws solid waste into a porous bag inside the solid waste container. Your urine, on the other hand (hopefully not literally, but this is microgravity after all and accidents do happen – talk to a cosmonaut if you don't believe me), is directed into the urine storage container, which activates the separator, fan, and the chemical pretreatment. Once combined with sulfuric acid and water, urine is directed into the dynamic gas-liquid separator. This is where the liquid and gas are separated. The liquid part is sent to the urine container, while the airflow is directed to the urine/water receptacle where a filter collects moisture from the air flow. Once odor has been taken out of the air flow, it is routed back to the module. In terms of actually operating the waste management system, it is not as easy or nearly as straightforward as using a terrestrial system.

For successful operation, the following steps must be taken:
— Check the toggle switch on the control panel is *on*.

To use the urine collector:
— Lift funnel from the holder.
— Remove the lid from funnel.

- Set the manual valve to the open position on the funnel.
- Verify the separator and dosage lights are on.
- Verify airflow.
- Hold funnel clear of the body.
- 20–30 sec. after finishing, close the manual valve on funnel.
- Verify the separator and dosage lights are *off*.
- Wipe funnel with washcloth and stow in the waste bag.
- Install lid on the funnel.
- Place funnel on holder.

For urination and defecation, the urine receptacle and solid/liquid waste collector must be used as follows:
- Remove lid from funnel.
- Without lifting funnel from holder, open plug valve.
- Verify separator light is *on*.
- Verify airflow.
- Prepare solid waste collector/insert collection bag.
- Lift funnel from holder.
- Use solid and liquid waste collector.
- Lift seat, remove collection bag, and stow in solid waste container.
- Wipe seat and funnel with washcloth/stow washcloth in waste bag.
- Place funnel on holder.
- Close lid on solid/liquid waste collector.
- Close manual valve on funnel.
- Verify separator light is *off*.
- Install lid on funnel.

Fire Detection and Suppression System

This system comprises one detector system for Zarya and one for Zvezda, a docking module detection system, fire extinguishers, and rebreather-style breathing apparatus. The "Signal" fire detection system, which comprises ten detectors (these are tested every 10 days by the crew), is designed to detect a particle count two times the reference value in the atmosphere. If the particle count is two to four times the reference value, a Class 2 *smoke* alarm is raised. If the particle count is four to ten times the reference value, a Class 1 *fire* alarm is raised. If this condition occurs, the caution and warning system lights up in red and is accompanied by a beeping sound. This status is automatically downlinked to mission control in Moscow. In the event of a fire, crews would don their rebreather-style breathing apparatus, which comprises a mask and a cartridge that contains an oxygen generating compound. The crewmember would then turn the starter lever on the mask by 180°, an action that would puncture an acid capsule, which would release its contents to react with aluminum oxide inside the briquette. Oxygen would then be released, and the crewmember could turn their attention to deploying their portable fire extinguishers, each containing 2.5 kilograms of fire extinguishing foam.

Each extinguisher when deployed is capable of 1-minute uninterrupted operation and can be deployed in either liquid spray or foam mode. If an open fire occurs, the liquid spray mode is used, whereas if only smoke is present, foam mode is used.

Paradigms in Life Support: The *Aggregation of Marginal Gains* and Thirsty Walls

» We searched for small improvements everywhere and found countless opportunities. Taken together, we felt they gave us a competitive advantage.
  Sir Dave Brailsford, Director, Sky (now INEOS) Professional Cycling Team

The Aggregation of Marginal Gains

You may be wondering what a quote from a director of a professional cycling team is doing in a textbook on life support systems, but I will explain. The *aggregation of marginal gains* is a phrase/concept attributed to Sky cycling coach Dave Brailsford, who not only led the British cycling team to ten gold medals at the Beijing Olympics but is also credited with helping the Sky professional cycling team (now INEOS) dominate the Tour de France in the 2010s.

The concept is very simple: by focusing on a 1 percent margin of improvement in every aspect of performance, it is possible to achieve a significant improvement when all those aggregations add up. For a professional cycling team, these marginal gains included implementing very precise food preparation procedures and having the cyclists bring their own mattresses and pillow with them when they traveled so they could get a good night's sleep before competition. In short, the concept is not a focus on perfection, but rather a focus on progression and *compounding improvements*.

A similar approach has been applied to spacecraft life support systems. Over many years of developing LSS technology, there have been no quantum leaps. Instead, LSS technology has developed by applying the marginal gain approach: filters have been refined, ventilation systems have been modified, and water recycling has been made more efficient. These are small improvements, but collectively they add up to a significant gain in terms of closing the LSS loop. As 2020 rolls into 2021, the LSS marginal gain approach continues, but there are some developments that promise more than a marginal gain. One of these developments applies to the air revitalization system and more specifically how the ARS system currently used on board submarines may be adjusted for spacecraft.

Thirsty Walls

When discussing what life support systems do well and what they don't do so well, one system that fits into the latter category is the CDRA. One question asked by students when I point out how difficult it is for the CDRA to do its job is this: Why

◘ Fig. 5.20 The officer of the watch control room aboard the USN Ohio Class Guided Missile Submarine USS Florida (SSGN 728) during operations in the Atlantic. Recent developments in life support technology may result in submarine carbon dioxide scrubbing technologies being used on spacecraft. Credit: USN

doesn't NASA simply use the carbon dioxide removal systems used on submarines? After all, submarines (◘ Fig. 5.20) carry over 100 crewmembers, and they don't seem to suffer from the effects of high carbon dioxide concentration.

As we know from our historical review of spacecraft life support systems, air revitalization is achieved by directing air through ducts and removal beds, which are packed with solids such as activated charcoal. But how does a submarine LSS air revitalization system work? These systems tend to be liquid-based capture systems, which are more compact, power efficient, and a whole lot more reliable than their space-based cousins. The problem until recently has been that these liquid-based capture systems need gravity. That's not to say such a system couldn't work in microgravity, but to do so would require a gas-permeable membrane, and the problem with membranes is that they suffer from slow kinetics and are susceptible to poisoning.

However, recent developments in additive manufacturing and capillary fluid mechanics now make it possible to expose liquids to cabin air in microgravity, which means that a reimagining of the spacecraft ARS is now possible. The name applied to this concept is *Thirsty Walls* (◘ Fig. 5.21). This approach may reduce the number of rotating elements in a traditional spacecraft ARS from 19 to 8 while also eliminating the high pressure and high flow elements. It would achieve this by using monoethanolamine (MEA), which is the liquid used to capture carbon dioxide on submarines [20]. Better still, there is some research suggesting that if MEA

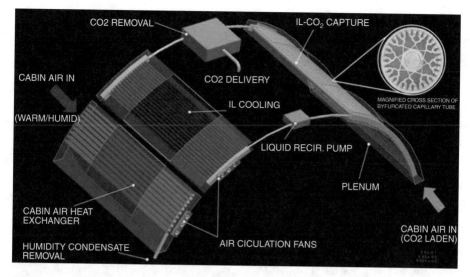

Fig. 5.21 The Thirsty Walls carbon dioxide removal concept. Credit: NASA

were replaced with ionic liquids, the ARS would be even more effective at scrubbing carbon dioxide than on submarines. If the system works, it would represent a transformational improvement in ARS technology on spacecraft – a quantum leap, as opposed to a marginal gain.

Key Terms

- Advanced Closed-Loop System (ACLS)
- Air Revitalization System (ARS)
- Carbon Dioxide Concentration Subsystem (CCA)
- Carbon Dioxide Reduction Assembly (CRA)
- Catalytic Oxidizer Assembly (COA)
- Charcoal Bed Assembly (CBA)
- Common Carrier Air Assembly (CCAA)
- Cooled Hydrophilic-Coated Heat Exchanger (CHX)
- Crew Quarters (CQ)
- Environmental Control and Life Support System (ECLSS)
- Environmental and Thermal Operating System (ETHOS)
- Fire Detection and Suppression System
- Fire Water Mist (FWM)
- Inter-module Ventilation (IMV)
- International Standard Payload Rack (ISPR)
- Life Support System (LSS)
- Lithium Hydroxide (LiOH)
- Monoethanolamine (MEA)
- Nitrogen Line Shut-Off Valve (NLSOV)

5

- Oxygen Generation Assembly (OGA)
- Oxygen Recharge Compressor Assembly (ORCA)
- Permanent Multipurpose Module (PMM)
- Portable Breathing Apparatus (PBA)
- Portable Fire Extinguisher (PFE)
- Russian Orbital Segment (ROS)
- Sorbent Bed Assembly (SBA)
- Temperature and Humidity Control (THC)
- Trace Contaminant Control System (TCCS)
- United States Orbital Segment (USOS)
- Urine Processing Assembly (UPA)
- Vacuum Dump Device (VADD)
- Vapor Compression Distillation (VCD)
- Venting Dump Device (VEDD)
- Venting Manual Return Valve (VEMRV)
- Volatile Organic Compound (VOC)
- Waste Line Shut-Off Valve (WLSOV)
- Water Processing Assembly (WPA)
- Water Pump Assembly (WPA)

❓ Review Questions

1. List the six subsystems of the USOS LSS.
2. What do the MPEVs do?
3. Explain how the CDRA scrubs carbon dioxide from the atmosphere.
4. What is the Sabatier reaction?
5. How does the TCCS remove contaminants from the atmosphere?
6. What is a VOC?
7. Describe the two ventilation processes in the USOS.
8. Explain the function of the VEDD, the WLSOV, the VEMRV, and the VADD.
9. List the five subsystems of the ROS LSS.
10. Explain the function of the Elektron unit.
11. How does the Vozdukh unit achieve its function?
12. What is the function of the Interface Pressure Integrity Monitors?
13. What is the difference between a Class 1 and a Class 2 alarm?
14. How might Thirsty Walls be used in a spacecraft LSS?

References

1. Wieland P O. (1994). *Designing for human presence in space: An introduction to environmental control and life support systems.* NASA RP-1324:5~48, 185 219~225.
2. Space Station Program Node Element 1 to Node Element 3 Interface Control Document, Part 1; SSP 41140, Revision E; August 24, 2005.
3. Node 1 Element to U.S. Laboratory Element Interface Control Document, Part 1; SSP 41141, Revision F; December 10, 2004.

4. Boeing – Huntington Beach. (1988, Feb 13). *International Space Station Node 1 pressure monitoring and equalization ECLSS verification analysis report*, MDC 96H0634B.
5. Boeing – Huntington Beach. (1988, Jan 30). *International Space Station Node 1 atmosphere control system distribution pressure loss ECLSS verification analysis report*, MDC 97H0374C.
6. Pressure Equalization of Vestibule with Node 1. Memorandum; 2-6920-ECLS-CHS-98-033; August 11, 1998. CONTACT David E. Williams NASA, Lyndon B. Johnson Space
7. Roy, R. J. Final report for the SPE Oxygen Generation Assembly (OGA); NASA Contract NAS8-38250-25; July 1995.
8. Gentry, G. J., Reysa, P. R., Williams, D. E. *International Space Station (ISS) Environmental Control and Life Support (ECLS) system overview of events: February 2004 – 2005. SAE Paper 2005- 01-2778*, 35th International Conference on Environmental Systems, July 12-15, 2005, Rome, Italy.
9. Williams, S. A. E. (2007). Paper 2007- 01-3099, 37th international conference on environmental systems, July 9-12, 2007, Chicago, Il.
10. Murdoch, K., Perry, J., Smith, F. (2003). Sabatier engineering development unit; SAE 2003-01-2496.
11. International Space Station (ISS) Environmental Control and Life Support (ECLS) System Overview of Events: February 2006 – 2007, Gregory J. Gentry, Richard P. Reysa, David E.
12. Zapata, J. L., & Chang, H. S. Analysis of air ventilation and crew comfort for the International Space Station Cupola. SAE 2002-01-2340.
13. Chang, H. S. Integrated computational fluid dynamics ventilation model for the International Space Station. SAE2005-01-2794.
14. Son, C. H., Barker, R. S., & McGraw, E. H. Numerical prediction and evaluation of space station intermodule ventilation and air distribution performance. SAE941509.
15. Abbud-Madrid, A., Amon, F. K., & McKinnon, J. T. *The MIST experiment on STS-107; fighting fire in microgravity*, 42nd AIAA Aerospace Sciences Meeting,1/5-8/2004, Reno Nevada.
16. Carriere, T., Butz, J., Naha, S., & Abbud-Madrid, A. *Fire suppression tests using a handheld water mist extinguisher designed for spacecraft application*, SUPDET 2012 Conference Proceedings, March 5–8, 2012, Phoenix, Arizona.
17. Portable Fire Extinguisher (FWM PFE), Revision A, September 2012.
18. Carter, D.L., Tobias, B., & Orozco, N. *Status of the regenerative ECLSS water recovery system*, AIAA 1278029, presented at the 42nd International Conference on Environmental Systems, San Diego, California, July, 2012.
19. Carter, D. L., & Orozco, N. *Status of the regenerative ECLSS water recovery system*, AIAA 1021863, presented at the 41st international conference on environmental systems, Portland, Oregon, July, 2011.
20. Dr. John Graf Dr. Mark Weislogel, Professor Thirsty Walls: A New Paradigm for Air Revitalization in Life Support. Final Technical Report of NIAC Phase 1 Study March 09, 2016 NASA Grant and Cooperative Agreement Number: NNH-15ZOA001N-15NIAC (Proposal #15-NIAC-15A0134) NIAC Phase I Study Period: June 2015 to March 2016.

Extravehicular Activity

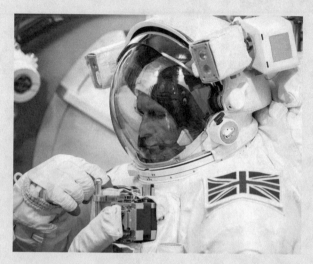

Credit: ESA

Contents

© Springer Nature Switzerland AG 2020
E. Seedhouse, *Life Support Systems for Humans in Space*,
https://doi.org/10.1007/978-3-030-52859-1_6

Learning Objectives
- Name the key elements that comprise the EMU
- Explain the function of the LCVG and LTA
- Describe how the EMU is donned
- Explain the rationale for the prebreathe
- Explain what is meant by denitrogenation
- Describe the ISLE protocol
- List four signs of DCS
- Explain how DCS is treated

Introduction

Without an extravehicular activity (EVA) capability, the ISS could not have been constructed. Period. Not only is EVA vital for the construction of orbiting outposts, but it also enables essential maintenance and repair activities to be conducted. EVA equipment – the *extravehicular mobility unit*, or EMU – is basically a life support suit that includes a portable life support system capable of providing its occupant with oxygen and removing carbon dioxide, just like the life support system on board the ISS. These EMUs are highly complex and extremely expensive (each one costs about $10 million), and their design crosses myriad engineering disciplines. Not surprisingly, training for an EVA is a time-consuming and exacting task, with at least 6 hours of practice required for every 1 hour of planned EVA. Then there is the issue of prebreathing, a carefully planned procedure that must precede each and every EVA. We'll learn about the prebreathe and all the other factors that comprise EVAs and EMUs in this chapter.

Extravehicular Activity Mobility Units

Extravehicular activity mobility units, or EMUs, provide astronauts with a pressurized environment, a supply of oxygen, removal of carbon dioxide, temperature and humidity control, and micrometeoroid/orbital debris (MMOD) protection. Components of the suit are available in different sizes, generally small, medium, and large, and are designed to fit male and female astronauts from the 5th to 95th percentile body size/height. A portable life support system (aka the backpack, aka PLSS) provides astronauts with life support system expendables, including oxygen, a battery, water for drinking and cooling, lithium hydroxide for removing carbon dioxide, and a 30-minute reserve of oxygen. The PLSS provides 7 hours of expendables as follows:
1. 15-min donning
2. 6 hours EVA
3. 15-min doffing
4. 30-min reserve/contingency

EMU Elements

An EMU (Fig. 6.1), which weighs 225 pounds and is pressurized to 4 psi, is designed for a 15-year life. It comprises a hard upper torso (HUT), lower torso assembly (LTA), liquid ventilation cooling garment (LVCG), gloves, helmet, communications carrier assembly (CCA), urine collection device (UCD), and bioinstrumentation system.

HUT and PLSS

The HUT provides the structural mounting for the helmet, arms, LTA, PLSS, display and control module (DCM), and electrical harness. The PLSS provides a structural mounting for oxygen bottles, water storage tanks, a fan, a separator and pump motor assembly, a sublimator, a contaminant control cartridge, regulators,

6

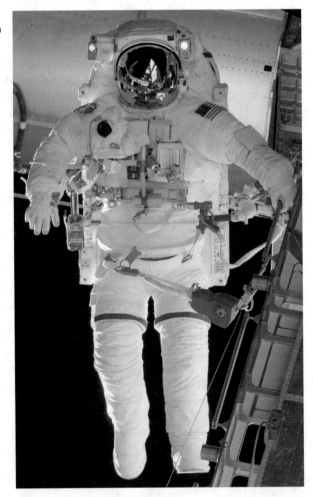

 Fig. 6.1   The EMU. Rick Mastracchio about to head out on an ISS construction EVA. During the 6-hour, 17-minute spacewalk, Mastracchio, together with Dave Williams, attached the Starboard 5 truss and retracted the forward heat-rejecting radiator from the station's Port 6 truss. Credit: NASA

valves, communications, and bioinstrumentation. Attached to the base of the PLSS is a secondary oxygen pack that contains 2.6 pounds of oxygen pressurized to 6000 psi. Other expendables contained within the PLSS include 1.2 pounds of oxygen stored at 850 psi, ten pounds of water for cooling, and lithium hydroxide, which is stored in the contaminant control cartridge.

LTA

The LTA comprises pants and knee and ankle joints and connects to the HUT using a waist ring. It is a multilayered system that includes a pressure bladder made of urethane-coated nylon, a restraining Dacron layer, an outer neoprene-coated nylon layer, four Mylar layers, and an outer layer of Goretex and Nomex. The gloves, which are available in 15 sizes, feature wrist connections, a wrist joint, and padding insulation for the palms and fingers.

Helmet

The helmet is made of polycarbonate that features a neck disconnect and ventilation pad. The assembly that slips over the helmet features various visors that can be adjusted by the astronaut to provide shielding from the sun, ultraviolet radiation, and micrometeoroids. Attached to either side of the helmet are two EVA lights (there is provision for mounting a video camera to the helmet too). Under the helmet is the Snoopy cap that features a microphone and headphones.

LVCG and CO_2

The LVCG, which weighs 6.5 pounds, has tubes integrated into the suit that circulate water to provide cooling. The tubing is 300 feet long and carries 240 pounds of water per hour through the system (◨ Fig. 6.2). The controls for the LVCG can be found on the DCM. Worn underneath the LVCG is the UCD, which can store about one quart of urine. Potable water is stored in the in-suit drink bag that can store 0.5 quarts of water. A tube from the HUT allows the astronaut to drink while suited. To remove carbon dioxide, the EMU relies on a cartridge comprising lithium hydroxide, charcoal, and filters. The cartridge is replaced following each EVA. Power for the whole kit and caboodle comes from silver-zinc rechargeable batteries.

Communication

Astronauts communicate with the ISS via the EVA communicator and EMY antenna. The communicator also telemeters the astronauts' ECG to the ISS. EVA radios, which weigh 8.7 pounds, have two single-UHF transmitters, three single-channel receivers, and a switching mechanism.

Instrumentation

The EMU electrical harness is integrated to the PLSS and provides biomedical instrumentation and connections for communications. Connected to the harness are the CCA and the bioinstrumentation system. Attached to the astronaut are

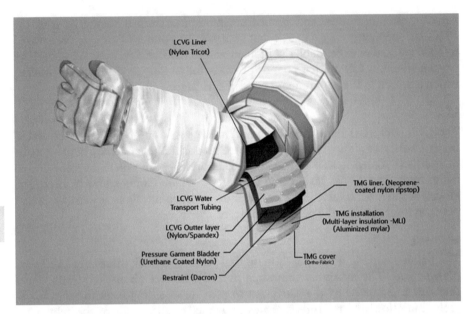

LCVG Liner
(Nylon Tricot)

TMG liner. (Neoprene-
coated nylon ripstop)

LCVG Water
Transport Tubing

TMG installation
(Multi-layer insulation -MLI)
(Aluminized mylar)

LCVG Outter layer
(Nylon/Spandex)

Pressure Garment Bladder
(Urethane Coated Nylon)

TMG cover
(Ortho-Fabric)

Restraint (Dacron)

Fig. 6.2 Cutaway of the EMU showing how the LCVG is integrated into the suit. Credit: NASA

ECG sensors, which are routed via the bioinstrumentation system to the communications system. Another instrumentation element is the DCM, which is attached to the HUT. This module features electrical controls, and a digital display of EMU controls that the astronauts can see while wearing the EMU. The DCM, which also enables astronauts to control the PLSS and the secondary oxygen pack, features a suit purge valve, a LCVG valve, and a control actuator for oxygen, which can be set to four positions. Additional controls include a voice communication switch, a power mode switch, and a digital display brightness control. Systems checks can also be scrolled through by the astronaut using the DCM, a feature that allows the astronaut to check the health of various systems such as the status of the primary oxygen tanks.

Oxygen

Oxygen enters the suit at the helmet. From there, it flows down the suit, and return air is directed through the contaminant control cartridge (CCC). Here, lithium hydroxide and charcoal remove carbon dioxide and odors. After passing through the CCC, oxygen is directed through a water separator which removes moisture. Oxygen is then directed through the fan before being routed through the sublimator, which cools the oxygen to 85°F. Finally, the oxygen passes through a vent and flow detector and is routed back into the suit. To protect the suit from excess pressure or oxygen depletion, the EMU features various sensors and the secondary oxygen pack which can maintain suit pressure at 3.45 psi.

Cooling Water System

This subsystem directs warm water from the LCVG and separates it into two loops. One loop is routed to the sublimator, where the water is cooled and returned to the cooling control valve. The second loop directs water to the cooling control valve directly. This is where the loops reunite, and full flow is channeled back into the LCVG. This arrangement ensures the LCVG has an uninterrupted flow of cooling water. During the process, the flow of water from the LCVG is first channeled through the gas separator, which removes gas from the loop, and then through a pump that regulates the flow at 260 pounds per hour. A side loop circulates 20 pounds per hour via the CCC to cool the lithium hydroxide canister.

PLSS Sensors

The PLSS is equipped with myriad sensors, including ones to detect air pressure, air flow, water pressure, water temperature, and carbon dioxide partial pressure. In addition to the sensors, the astronauts have various valves available that allow them to purge oxygen, change the cooling, check oxygen supply, and check pressure.

Donning the EMU

Prior to donning the EMU, astronauts must first begin the process of prebreathing, an explanation of which follows this section. Donning the EMU (see Appendix V for a more detailed explanation) can best be explained in a series of 22 steps as follows:

1. Don the MAG
2. Don the LCVG
3. Attach the EEH to the HUT
4. Attach the DCM to the HUT (PLSS is pre-attached to the HUT)
5. Attach the arms to the HUT
6. Rub helmet with anti-fog solution
7. Attach wrist mirror and checklist on the sleeves
8. Insert a food bar and water-filled IDB inside the HUT
9. Check lights and TV cameras on the EVA
10. Place EVA over the helmet
11. Connect CCA to the EEH
12. Step into LTA and pull it above waist
13. Plug SCU into DCM
14. Wriggle into HUT
15. Attach the cooling tubes of the LVCG to the PLSS
16. Attach the EEH electrical connections to the PLSS
17. Lock the LTA to the HUT
18. Put on the CCA
19. Put on comfort gloves

20. Lock on helmet and EVA
21. Lock outer gloves
22. Check EMU for leaks by increasing the pressure to 0.20 atm above airlock pressure

Extravehicular Activity Prebreathe Protocols

Crews inside the ISS breathe air at a pressure of 14.7 pounds per square inch (psi), but the pressure inside the EMU is 4.3 psi. The lower pressure is necessary because operating a spacesuit at the same pressure as inside the ISS would result in a large pressure differential between the EMU and the vacuum of space. So, this means the pressure inside the EMU must be reduced. By reducing suit pressure, spacewalking astronauts find it easier to move around, but there is a disadvantage of operating at a lower pressure. The disadvantage is that when transitioning from a high to a low pressure, there is a risk of nitrogen bubbles forming in the blood. These bubbles can cause decompression sickness or DCS. In an effort to prevent DCS, various prebreathe protocols have been developed. Some of these protocols were performed at rest, others used exercise to accelerate nitrogen removal (a procedure termed *denitrogenation*), and yet others applied staged protocols [1, 2, 3]. One such protocol is the in-suit light exercise (ISLE) protocol [4, 5, 6]. To date, this protocol and others that preceded it have been effective in avoiding the occurrence of DCS. Having said that, although there have been no DCS incidences, there are procedures in place in the event that an astronaut suffers from decompression sickness. These procedures and select protocols developed for avoiding DCS will be discussed here.

We'll begin with the risk of nitrogen bubbles forming the blood. Under normal atmospheric conditions, nitrogen is dissolved in the blood and tissues. But when pressure is reduced quickly, nitrogen comes out of solution and forms bubbles. An analogy to this is when opening a can of Coke – as soon as the can is opened, the dissolved carbon dioxide instantly forms bubbles due to the reduction in pressure. In the body, nitrogen bubbles released into the bloodstream can lead to a constellation of symptoms ranging from numbness to joint pain to death (❏ Table 6.1). The term applied to describe bubble formation in the tissues and blood caused by a reduction in environmental pressure is *venous gas emboli* (VGE) [7, 8, 9, 10, 11, 12], which can be detected using ultrasound imaging.

A high number of VGE results in *decompression sickness (DCS)*, also known as "the bends" [13, 14, 15, 16, 17, 18]. During the Shuttle era, DCS risk was reduced by implementing an oxygen prebreathe protocol that washed out tissue nitrogen before EVAs. But during the ISS era, with oxygen being a limited resource, it was necessary to develop a protocol that reduced the amount of oxygen used. One such protocol was the aforementioned ISLE, which is discussed later. But first, a brief overview of the history of the prebreathe protocol.

■ **Table 6.1**   Symptoms of decompression sickness, listed from most commonly to least commonly reported

Pain, especially near joints	Muscles discomfort
Numbness or paresthesia	Impaired cognitive function
Headache, light-headedness, and fatigue	Pulmonary problems ("the chokes")
Malaise, nausea, vomiting	Impaired coordination
Dizziness, vertigo	Reduced consciousness
Motor weakness	Lymphatic concerns such as swelling
Skin mottling	Compromised cardiovascular function

History of Prebreathe Protocols

Prebreathe (PB) protocols have been developed since the 1970s. Some used mathematical modeling, some used simulations, and some used human testing [19, 20, 21, 22, 23], but one common denominator applied to protocol development (■ Table 6.2) was designing the protocol with operational issues in mind. During the ISS era, the focus on operations has not changed, but as more data was generated, the prebreathe protocol evolved. Early in the evolution of protocol development, prebreathe protocols lasted up to 4 hours. Then exercise was added as a variable [14, 24, 25, 26, 27], and the length of protocols was reduced. The reason for adding exercise was that it accelerated the rate of denitrogenation. In addition to including exercise to accelerate denitrogenation, staged decompression protocols were developed until eventually a 2-hour exercise prebreathe protocol – the ISLE – was developed for ISS.

Select Prebreathe Protocols

To date, prebreathe protocols have been employed very successfully, with no reported cases of DCS. A major reason for this safety record has been the robustness of experimental ground-based protocols that have involved hundreds of tests conducted by hundreds of subjects. One of the major changes to the protocol occurred during the transition from the Shuttle Program to ISS operations. Astronauts performing EVAs from the Shuttle had to undergo a lengthy decompression protocol, which reduced mission efficiency [25, 28]. With construction of the ISS requiring more than 160 EVAs, NASA had to design a more efficient decompression protocol, some of which are reviewed here.

6

◼ **Table 6.2**	Prebreathe Protocols
(a) Campout prebreathe protocol proposed 08/16/04	
1	30-min oxygen prebreathe
2	31-min oxygen depressurization from 14.7 to 10.2 psi
3	8 hours 40 min at 10.2 psi/26.5% oxygen
4	10-min repressurization to 14.7 psi on oxygen prebreathe
5	30-min hygiene break while still on oxygen prebreathe
6	31-min oxygen depressurization to 10.2 psi
7	60-min suit donning at 10.2 psi while on 26.5% oxygen
8	17-min purge and leak check
9	40-min oxygen in-suit prebreathe
10	10-min additional in-suit prebreathe
11	30-min depressurization to 4.3 psi
(b) Exercise prebreathe protocol as flown on ISS	
1	10-min dual-cycle ergometer at 75% of Vo2 for last 7 min
2	24-min of intermittent exercise starting 55 min into PB and ending 95 min
3	30-min ascent to 10.2 psi on 100% oxygen
4	30-min at 10.2 psi breathing 73.5% N2 and 26.5% oxygen
5	17-min purge and leak check
6	5-min on 100% oxygen then descent to 14.7 psi
7	35-min in-suit PB
8	20-min additional in-suit PB to compensate for no in-suit Doppler
9	30-min ascent to 4.3 psi

The In-Suit Light Exercise Protocol

After testing various protocols beginning in 1997, NASA eventually zeroed in on a promising protocol known as the Phase V-5 Protocol. The testing for this protocol comprised a multicenter trial that included test participants representative of the astronaut population with respect to age, gender, fitness, and percentage body fat. The final protocol consists of the steps outlined below.

1. 60 minutes of exercise breathing oxygen while simultaneously preparing for the EVA, followed by depressurization to 10.2 psi while conducting light exercise (at 5.8 ml per kilogram of body weight per minute of exercise) breathing enriched air.
2. 30-minute suit donning at 10.2 psi.

3. 50 minutes in-suit light activity at 6.8 ml per kilogram of body weight per minute of exercise – equivalent to walking a mile in 70 minutes – while breathing oxygen.
4. 50 minutes in-suit prebreathe at rest, breathing oxygen.

The ISLE (◘ Fig. 6.3) has its advantages and disadvantages. One disadvantage is a longer in-suit prebreathe and the challenges of monitoring metabolism, but this is outweighed by the advantages, one of which is saving 2.5 kilograms of oxygen per EVA compared with the Shuttle protocol. This advantage is a significant one because oxygen is a much more limited resource on the ISS than on the Shuttle. A second advantage is the reduced time astronauts have to spend with masks on (◘ Fig. 6.4), and a third is not having the astronauts isolated in the airlock.

Rest	Light Exercise (EVA Prep) (5.8 mL*kg⁻¹*min⁻¹)		Prebreathe Light In-Suit Exercise (6.8 mL*kg⁻¹*min⁻¹)	Rest	Flight Simulation Ascent Rest	30,300 ft Light Exercise
		Depress to 10.2 psi	Repr			
	40 min	20 min	5	45 min		
		60 min		30 min	50 min	
130 min			190 min		50 min	30 min	240 min
Air	Oxygen		0.265 Oxygen		Oxygen	

◘ **Fig. 6.3** Timeline of the ISLE prebreathe protocol. Credit: NASA

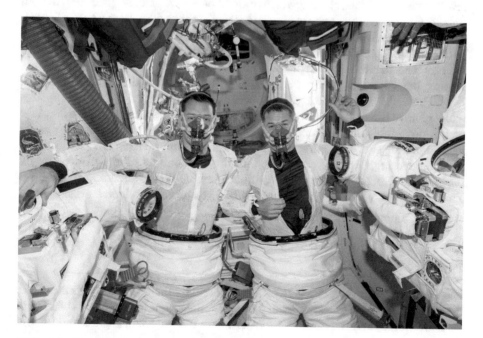

◘ **Fig. 6.4** Shane Kimbrough (R) and Thomas Pesquet (L) conduct a prebreathe inside Quest, the ISS module that includes the airlock and equipment lock (where the spacesuits are stored). During their EVA, the astronauts installed new adapter plates and hooked up electrical connections for new lithium-ion batteries. Credit: NASA

Campout Procedure

A second protocol, also used on the ISS, is the *campout protocol*, which has the two EVA astronauts "camp out" in the airlock at 10.2 psi, with an oxygen concentration of 26.5 % during the night preceding the spacewalk. The time spent at 10.2 psi is 8 hours and 40 minutes, and most of this time is spent sleeping. After waking, the astronauts are repressurized to 14.7 psi while breathing 100% using a mask. They breathe oxygen for 70 minutes, during which time they eat a snack and use the restroom. Then they resume breathing 26.5% oxygen at 10.2 psi, masks are removed, and the astronauts begin donning their suits. After the suits are donned, the airlock is repressurized to 14.7 psi, and a third astronaut enters the airlock to assist with the prebreathe before final depressurization to the vacuum of space. This final phase lasts 50 minutes.

Cycle Ergometer with Vibration Isolation and Stabilization Protocol

A third protocol is the *cycle ergometer with vibration isolation and stabilization (CEVIS) protocol*. Prior to launch, astronauts who will be using the CEVIS protocol perform a maximum oxygen uptake test using leg ergometry. Based on the results of this test, an exercise protocol is devised that distributes the workload between 12% upper body and 88% lower body. When preparing for the EVA, the astronauts breathe oxygen via a mask and perform 3 minutes of exercise at 75 revolutions per minute, starting at a workload of 37.5% of their maximum oxygen uptake. Gradually, this exercise intensity is increased to 50%, then 62.5% of maximal oxygen uptake, and finally 7 minutes at 75% of their maximal oxygen uptake. After completing their exercise, the astronauts breathe pure oxygen for another 50 minutes, and the airlock is depressurized to 10.2 psi over a period of 30 minutes. During the depressurization period, the astronauts don their LCVG and the LTA. Then, once the airlock oxygen concentration is stabilized at 26.5%, the astronauts remove their masks and complete donning of the spacesuit. The advantage of this protocol is that for much of the PB phase, astronauts are actively engaged in donning the suit. Following donning, the astronauts perform a leak check and then purge using 100% oxygen. Once this phase is complete, the astronauts begin the in-suit PB phase, which begins with a 5-minute depressurization to 14.7 psi. This phase, which lasts for 55 minutes, is followed by airlock depressurization and suit depressurization to 4.3 psi.

Airlock

On board the ISS, the place where astronauts conduct their prebreathe, is the Quest module (◘ Fig. 6.5). Quest is divided into the equipment lock, which is where the suits are stored, and the airlock, which is where the astronauts egress and ingress the station.

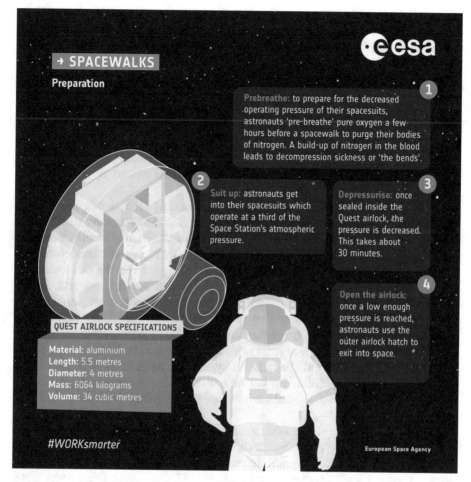

□ **Fig. 6.5** A review of the prebreathe and the dimensions of Quest. Credit: ESA

Treatment of Decompression Sickness

Although there have been no incidences of DCS on board the ISS, there are procedures in place in the event that an astronaut suffers the bends [29]. DCS treatment on orbit is similar to treatment on Earth, requiring the administration of fluids and hyperbaric oxygen depending on the severity of symptoms. A limited neurological examination is available during the course of DCS treatment. If an astronaut was diagnosed with DCS on ISS today, they would be repressurized to maximum cabin atmosphere immediately while continuing to breathe 100% oxygen. If symptoms did not resolve, the Bends Treatment Adapter (BTA) would be installed, and a maximum suit overpressure of 8 psi could be applied so the astronaut would be breathing at 24 psi (subjecting the suit to this pressure would run the risk of grounding the suit for operational use). At this pressure, there would be more than an 80%

decrease in the volume of gas, and symptoms would be expected to subside. Note: during the Shuttle era, there was an option to return to Earth, but that option is not available to ISS crews and nor will it be for lunar or Mars crews in the future.

Ebullism

Ebullism, an occupational hazard among spacefarers, is a term that describes the outcome of exposure of the body to vacuum [30, 31]. Such an event, which could occur during a *severe* EVA mishap, would result in profound and life-threatening physiological consequences that probably couldn't be treated by an on board recompression facility.

So, what happens to a human exposed to a vacuum? First, damage to the pulmonary tissue would be catastrophic due to dramatic overpressure caused by the extraordinary pressure differential [32]. This pressure differential would cause tearing of the pulmonary tissue, alveoli rupturing, hemorrhaging, and atelectasis [33]. While this damage is being inflicted on the pulmonary system, the cardiovascular system would fare little better. The myocardium would be stretched, and cardiac contractility would only be maintained for a 5 or 6 minutes at most. Initially, due to anoxia [30, 34], the heart would try to compensate by increasing heart rate for about 15 to 20 seconds, but heart rate would fall below baseline within just one minute. One minute later, the arterial pressure wave would be lost. Damage to the central nervous system would be inflicted by severely reduced cerebral blood flow and cerebral anoxia. Additionally, ischemia and thrombosis caused by the blockage of blood flow at the blood-bubble interface would quickly inflict impaired cognitive function.

While this constellation of symptoms suggests ebullism would probably be a fatal event, the reality may be slightly different. For example, primate studies conducted during the Apollo Program that exposed research subjects to 120,000-foot altitude revealed a 94% rate of survival [35, 36, 37, 38]. Given this survival rate, treatment protocols could be developed. Currently there are no such protocols, but research suggests there is a way to save an astronaut exposed to such an event. The first step would be immediate return to pressure since this would reverse the VGE and reduce tissue swelling which in turn would permit further treatment. Once the patient had been returned to the spacecraft habitat, hyperbaric oxygen therapy could be applied. Research has shown that pressure as high as six atmospheres would probably need to be used since a 100% oxygen atmosphere could not be used due to toxicity. The next step would necessitate endotracheal suctioning and intubation in an effort to prevent pulmonary barotrauma and cardiac compromise. The internal bleeding the patient would have suffered could be treated by the use of fluid expanders such as Dextran. Another step in treatment could be the use of a defibrillator in an attempt to stabilize cardiac contractility. Finally, a drug treatment designed to counter peripheral vascular insufficiency and cerebral hypoxia and prevent calcium loading would be required.

Summary

DCS is an occupational hazard for those who venture into space. To date, this mission risk has been mitigated effectively by applying rigorously tested prebreathe protocols that have been validated specifically for spaceflight. But planned missions beyond Earth orbit will require the development and validation of enhanced prebreathe protocols and research into the viability of using variable-pressure EVA suits as a means of mitigating the DCS risk.

Key Terms
- Bends Treatment Adapter (BTA)
- Common Carrier Assembly (CCA)
- Contaminant Control (CCC)
- Cycle Ergometer with Vibration Isolation and Stabilization (CEVIS)
- Decompression Sickness (DCS)
- Display Control Module (DCM)
- Extravehicular Activity (EVA)
- Extravehicular Activity Unit (EMU)
- Hard Upper Torso (HUT)
- Liquid Cooling Ventilation Garment (LCVG)
- Lower Torso Assembly (LTA)
- Micrometeoroid Orbital Debris (MMOD)
- Portable Life Support System (PLSS)
- Urine Collection Device (UCD)
- Venous Gas Emboli (VGE)

? **Review Questions**
1. List any six elements that comprise the EMU.
2. What is the function of the LCVG?
3. List four functions of the PLSS.
4. What is the CCC used for?
5. What are VGE and how can they be detected?
6. What is meant by the term denitrogenation?
7. List six symptoms of DCS.
8. What are the four stages of the ISLE?

References

1. Conkin, J., Gernhardt, M. L., Powell, M. R., & Pollock, N. (2004). A probability model of decompression sickness at 4.3 psia after exercise prebreathe. In *NASA Technical Publication NASA/TP-2004-213158*. Houston: Johnson Space Center.
2. Dixon, G. A., Adams, J. D., Olson, R. M. &, Fitzpatrick, E. L. (1980). Validation of additional prebreathing times for air interruptions in the shuttle EVA prebreathing profile. In Proceedings of the 1980 Aerospace Medical Association Annual Scientific Meeting, Anaheim, CA, 16–7.
3. Gernhardt, M. L., Conkin, J., Foster, P. P., Pilmanis, A. A., Butler, B. D., Fife, C. E., et al. (2000). Design of a 2-hr prebreathe protocol for space walks from the international space station. [Abstract # 43]. *Aviation, Space, and Environmental Medicine, 71*, 49.
4. Gernhardt, M. L., Dervay, J. P., Welch, J., Conkin, J., Acock, K., Lee, S., Moore, A., & Foster, P. (2003). Implementation of an exercise prebreathe protocol for construction and maintenance of the international space station- results to date. [Abstract # 145]. *Aviation, Space, and Environmental Medicine, 74*, 397.
5. Kumar, K. V., Waligora, J. W., & Gilbert, J. H., III. (1992). The influence of prior exercise at anaerobic threshold on decompression sickness. *Aviation, Space, and Environmental Medicine, 63*, 899–904.
6. Webb, J. T., Fischer, M. D., Heaps, C. L., & Pilmanis, A. A. (1996). Exercise-enhanced preoxygenation increases protection from decompression sickness. *Aviation, Space, and Environmental Medicine, 67*, 618–624. Webb JT, Kannan N, Pilmanis AA. Gender not a factor for altitude decompression sickness risk. Aviation, Space, and Environmental Medicine. 2003; 74:2-10.
7. Blatteau, J.-E., Souraud, J.-B., Gempp, E., & Boussuges, A. (2006). Gas nuclei, their origin, and their role in bubble formation. *Aviation, Space, and Environmental Medicine, 77*, 1068–1076.
8. Boothby, W. M., Luft, U. C., & Benson, O. O., Jr. (1952). Gaseous nitrogen elimination. Experiments when breathing oxygen at rest and at work with comments on dysbarism. *Journal of Aviation Medicine, 23*, 141–176.
9. Foster, P. P., Feiveson, A. H., Glowinski, R., Izygon, M., & Boriek, A. M. (2000b). A model for influence of exercise on formation and growth of tissue bubbles during altitude decompression. *American Journal of Physiology-Regulatory Integrative and Comparative Physiology, 279*, R2304–R2316.
10. Cameron, B. A., Olstad, C. S., Clark, J. M., Gelfand, R., Ochroch, E. A., & Eckenhoff, R. G. (2007). Risk factors for venous gas emboli after decompression from prolonged hyperbaric exposures. *Aviation, Space, and Environmental Medicine, 78*, 493–499.
11. Kumar, K. V., Powell, M. R., & Waligora, J. M. (1993a). Evaluation of the risk of circulating microbubbles under simulated extravehicular activity after bed rest. In SAE Technical Series No. 932220. 23rd International Conference on Environmental Systems. Colorado Springs, CO, 5.
12. Powell, M. R., Waligora, J. M., Norfleet, W. T., & Kumar, K. V. (1993). *Project ARGO - Gas phase formation in simulated microgravity. NASA Technical Memorandum 104762*. Johnson Space Center: Houston.
13. Conkin, J., Powell, M. R., Foster, P. P., & Waligora, J. M. (1998). Information about venous gas emboli improves prediction of hypobaric decompression sickness. *Aviation, Space, and Environmental Medicine, 69*, 8.
14. Conkin, J., Waligora, J. M., Horrigan, D. J., Jr., & Hadley, A. T., III. (1987). *The effect of exercise on venous gas emboli and decompression sickness in human subjects at 4.3 psia. NASA Technical Memorandum 58278*. Johnson Space Center: Houston.
15. Horrigan, D. J., & Waligora, J. M. The development of effective procedures for the protection of space shuttle crews against decompression sickness during extravehicular activities. Proceedings of the 1980 Aerospace Medical Association Annual Scientific Meeting, Anaheim, CA, May, 1980; 14-5. Risk of Decompression Sickness (DCS) 65
16. Pilmanis, A. A., Petropoulos, L. J., Kannan, N., & Webb, J. T. (2004). Decompression sickness risk model: development and validation by 150 prospective hypobaric exposures. *Aviation, Space, and Environmental Medicine, 75*, 749–759.

17. Ryles, M. T., & Pilmanis, A. A. (1996). The initial signs and symptoms of altitude decompression sickness. *Aviation, Space, and Environmental Medicine, 67*, 983–989.
18. Dixon, J. P. (1992). Death from altitude-induced decompression sickness: major pathophysiologic factors. In A. A. Pilmanis (Ed.), *The Proceedings of the 1990 Hypobaric Decompression Sickness Workshop* (Report AL-SR-1992-0005) (pp. 97–105). San Antonio: Brooks AFB.
19. Dervay, J., & Gernhardt, M. (2001). Decompression sickness in spaceflight: Likelihood, prevention and treatment. Version 1.04.
20. Hankins, T. C., Webb, J. T., Neddo, G. C., Pilmanis, A. A., & Mehm, W. J. (2000). Test and evaluation of exerciseenhanced preoxygenation in U-2 operations. *Aviation, Space, and Environmental Medicine, 71*, 822–826.
21. Loftin, K. C., Conkin, J., & Powell, M. R. (1997). Modeling the effects of exercise during 100% oxygen prebreathe on the risk of hypobaric decompression sickness. *Aviation, Space, and Environmental Medicine, 68*, 199–204.
22. Conkin, J., Kumar, K. V., Powell, M. R., Foster, P. P., & Waligora, J. M. (1996). A probabilistic model of hypobaric decompression sickness based on 66 chamber tests. *Aviation, Space, and Environmental Medicine, 67*, 176–183.
23. Webb, J. T., & Krutz, R. W. (1988). An annotated bibliography of hypobaric decompression sickness research conducted at the crew technology division, USAF School of Aerospace Medicine, Brooks AFB, Texas from 1983-1988. USAFSAM-TP-88-10, Brooks AFB, TX
24. Dujić, Z., Duplancic, D., Marinovic-Terzic, I., Bakovic, D., Ivancev, V., Valic, Z., Eterovic, D., Petri, N. M., Wisløff, U., & Brubakk, A. O. (2004). Aerobic exercise before diving reduces venous gas bubble formation in humans. *Journal of Physiology, 555*, 637–642.
25. Gernhardt, M. L. Overview of Shuttle and ISS Exercise Prebreathe Protocols and ISS Protocol Accept/Reject Limits. Prebreathe Protocol for Extravehicular Activity Technical Consultation Report; 96-125; NASA/TM-2008-215124
26. Krutz, R. W., & Dixon, G. A. (1987). The effect of exercise on bubble formation and bends susceptibility at 9,100 m (30,000 ft; 4.3 psia). *Aviation, Space, and Environmental Medicine, 58*(9, Suppl), A97–A99.
27. Pollock, N. W., Natoli, M. J., Vann, R. D., Nishi, R. Y., Sullivan, P. J., Gernhardt, M. L., Conkin, J., & Acock, K. E. (2004a). High altitude DCS risk is greater for low fit individuals completing oxygen prebreathe based on relative intensity exercise prescriptions. [Abstract #50]. *Aviation, Space, and Environmental Medicine, 75*, B11.
28. McIver, R. G., Beard, S. E., Bancroft, R. W., & Allen, T. H. (1967). Treatment of decompression sickness in simulated space flight. *Aerospace Medicine, 38*, 1034–1036.
29. Pilmanis, A. A., Webb, J. T., Balldin, U. I., Conkin, J., & Fischer, J. R. (2010). Air break during preoxygenation and risk of altitude decompression sickness. *Aviation, Space, and Environmental Medicine, 81*, 944–950.
30. Rudge, F. W. (1992). The role of ground level oxygen in the treatment of altitude chamber decompression sickness. *Aviation, Space, and Environmental Medicine, 63*, 1102–1105.
31. Vann, R. D., Gerth, W. A., Leatherman, N. E., & Feezor, M. D. (1987). A likelihood analysis of experiments to test altitude decompression protocols for shuttle operations. *Aviation, Space, and Environmental Medicine, 58*, A106–A109.
32. Conkin, J., Edwards, B., Waligora, J., & Horrigan, D. (1987). Empirical Models for Use in Designing Decompression Procedures for Space Operations. NASA-TM-100456, 1–52
33. Hall, W. M., & Cory, E. L. (1950). Anoxia in Explosive Decompression Injury. *American Journal of Physiology, 160*, 361–365.
34. Dunn, J. E., Bancroft, R. W., Haymaker, W., & Foft, D. W. (1965). Experimental Animal Decompressions to Less Than 2 mmHg Abs. (Pathological Effects). *Aerospace Medicine, 36*, 725–732.
35. Burch, B. H., Kemp, J. P., Vail, E. G., Frye, S. A., & Hitchcock, F. A. (1952). Some Effects of Explosive Decompression and Subsequent Exposure to 30 mmHg Upon the Hearts of Dogs. *Journal of Aviation Medicine, 23*, 159–167.

36. Cooke, J. P., & Bancroft, R. W. (1966). Some Cardiovascular Responses in Anesthetized Dogs During Repeated Decompressions to a Near-Vacuum. *Aerospace Medicine, 37,* 1148–1152.
37. Edelmann, A., Whitehorn, W. V., Lein, A., & Hitchcock, F. A. Pathological Lesions Produced by Explosive Decompression, WADC-TR-51-191.
38. Koestler, A. G. (1967). Replication and Extension of Rapid Decompression of Chimpanzees to a Near Vacuum. ARL-TR-67-2, Aeromedical Research Lab, Holloman Air Force Base.

Suggested Reading

Conkin, J., Gernhardt, M. L., Powell, M. R., & Pollock N. (2004). A probability model of decompression sickness at 4.3 psia after exercise prebreathe. NASA Technical Publication NASA/TP-2004-213158, Houston: Johnson Space Center

Jenkins, D. R. Dressing for Altitude: U.S. Aviation Pressure Suits, Wiley Post to Space Shuttle. NASA SP; 2011-595). ISBN 978-0-16-090110-2. Also available as a free online book at: https://www.nasa.gov/pdf/683215main_DressingAltitude-ebook.pdf.

Thomas, K. S., & McMann, H. J. (2006). *US spacesuits*. Springer-Praxis.

6

Countermeasures

Credit: NASA

Contents

© Springer Nature Switzerland AG 2020
E. Seedhouse, *Life Support Systems for Humans in Space*,
https://doi.org/10.1007/978-3-030-52859-1_7

🎓 **Learning Objectives**

After reading this chapter, you should be able to:

- Describe the history of exercise countermeasures in space
- Describe the characteristics of the USOS Countermeasure Program
- Explain what happens to exercise capacity during long duration spaceflight
- Explain what is meant by the term *permissible exposure limits* in the context of radiation exposure
- Explain what is meant by the ALARA Principle
- Explain what is meant by the AHARS Principle
- Describe how active and passive dosimeters are used on the ISS
- List four detectors used on ISS today and describe how each functions
- Explain how the MATROSHKA experiment has helped estimate radiation exposure
- Explain the utility of water as a radiation shield
- Explain what is meant by LET and RBE
- Explain the mechanism of radiation injury and repair
- List three potential pharmacological radiation countermeasures and explain how they work
- Describe how immune system function is suppressed in spaceflight
- Describe how supplements can boost immune system function

Introduction

Spending time in space results in a number of physiological changes that may exert a profound negative effect on astronaut health. These changes include a reduction in maximum oxygen uptake, a reduction in muscle size and strength, and a reduction in bone mineral density (BMD). One way of managing these effects is by the rigorous application of countermeasures such as exercise. During the International Space Station (ISS) era, astronauts have had a choice of exercise equipment, including the Advanced Resistive Exercise Device (ARED), the cycle ergometer, the Flywheel Exercise Device (FWED), and the inventively named Combined Operational Load Bearing Resistance Treadmill (COLBERT, after talk show host Stephen Colbert).

But as space agencies turn their attention to missions beyond Earth orbit, these countermeasures will assume ever more importance due to the sheer length of the missions. We already know that astronauts lose 1.0 to 1.2 percent of their BMD every month on orbit. While that is a significant amount of the bone, it is to a certain extent manageable, because astronauts spend only 6 months in LEO before they return to the comforts of their respective agencies and the care of the rehabilitation specialists.

So what will happen during (and following) a Mars mission? First there is the 6-month outbound journey, during which astronauts will lose about seven percent of their BMD. Then there is the surface stay. Now, the first mission will most likely feature just 1 month on the surface and because Mars has a reduced gravity, astronauts won't lose quite as much bone, although they can still expect to lose about

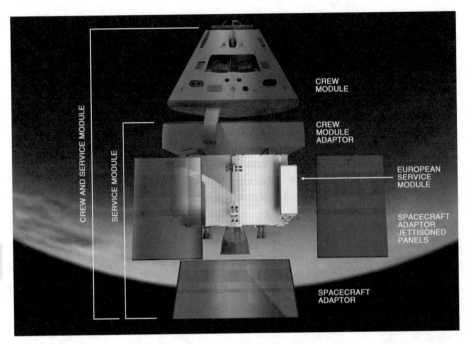

◘ Fig. 7.1 Schematic of NASA's Multi-purpose Crew Vehicle (MPCV), also known as Orion. Credit: NASA

another half a percent of their BMD. But during the trip back to Earth, each crew-member may lose another seven percent of their BMD (or more – we just don't know), which brings the total to about 15%. That could be catastrophic. Hence the need for countermeasures. But compared with the roomy ISS, NASA's spacecraft (◘ Fig. 7.1) is tiny, which means space for countermeasures will be at a premium.

We'll return to the subject of Mars-bound countermeasures later, but first, it is instructive to take a look at the history of the use of exercise in space.

A Short History of the Application of Exercise in Space

For those interested in this subject, I refer you to Moore [1], who has written comprehensive accounts of the development of exercise countermeasures, and Hackney [2], who has compiled a succinct summary of the exercise devices employed in each program. The first application of exercise in the US space program was in Project Mercury, when astronaut candidates were subjected to various exercise tests as part of astronaut selection. Exercise was also a component of cosmonaut selection during this era as depicted in ◘ Fig. 7.2.

Despite the short Mercury missions, there were some postflight reports of postural hypotension following landing, which prompted NASA's Medical Operations Office to note that a "prescribed inflight exercise program may be necessary to preclude symptoms in case of the need for an emergency egress soon after landing"

Fig. 7.2 Yuri Gagarin undergoes exercise testing. Credit: Russian Space Agency

[3]. In the Gemini Program, more extensive physiological testing was conducted. And, since these missions were much longer, it wasn't surprising that physiological effects were more pronounced. For example, following the Gemini VII flight, it was discovered that 14 days spent in space was sufficient to significantly reduce exercise capacity. As a consequence of the biomedical results of the Gemini Program, it was deemed prudent to implement inflight exercise as a countermeasure during the Apollo Program. So, astronauts used a modified off-the-shelf inflight exercise device (**Fig. 7.3**). Although Apollo astronauts became the first to use exercise as a countermeasure, the physiological benefits of the training were unclear, although the astronauts did report that exercise helped them relax.

The next program was Skylab. Skylab hosted three crews and the length of the missions (SL-2, 28 days; SL-3, 56 days; SL-4, 84 days) enabled a veritable treasure trove of biomedical data to be collected. During SL-2, astronauts were allocated 30 minutes of exercise per day using the cycle ergometer (**Fig. 7.4**). The time was increased to 60 minutes per day during Skylab 3 and to 90 minutes per day during Skylab 4.

◘ Fig. 7.3 Spending a long time in space can cause muscle atrophy and loss of bone mass. To ameliorate these effects, astronauts used an inflight exercise device that permitted isometric and isotonic exercise. Credit: NASA

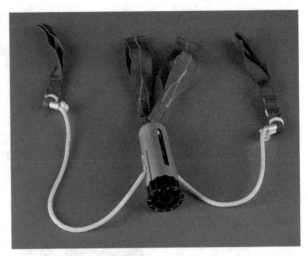

7

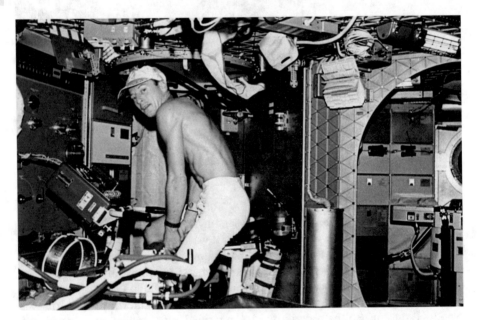

◘ Fig. 7.4 Charles Conrad exercises on the cycle ergometer during Skylab 2. The ergometer was used to conduct the vectorcardiogram experiment MO93 which assessed changes in the astronauts' cardiovascular systems. Credit: NASA

The physiological data collected during Skylab was extensive and can be found in the following 230 MB online document: ► https://ntrs.nasa.gov/archive/nasa/casi.ntrs.nasa.gov/19770026836.pdf. What follows is a (very!) brief snapshot of those findings. In the Skylab-2 crew, leg extensor strength was reduced by ~25% (at a rate of 0.9% per day) and by ~10% (at a rate of 0.1% per day) in the Skylab-4 crew. After landing, crews were able to stand and walk without difficulty. Exercise capacity across all three missions seemed to suffer little to no decrement, despite the limitations in ergometer capability. Perhaps thanks to the increased amount of time spent exercising during Skylab-4, this crew's cardiovascular parameters recovered quicker than the preceding two crews.

Thanks to the Skylab data, exercise countermeasures were more defined at the beginning of the Shuttle era which spanned 30 years and 135 flights. Missions ranged from just 2 days to more than 17 days. The primary exercise device was the cycle ergometer (◘ Fig. 7.5), although treadmills and a rowing ergometer were also evaluated during the program. Shuttle flight rules required that exercise be performed every other day by the Commander, Pilot, and Flight Engineer and every third day for Mission Specialists, although the intensity and duration were not prescribed [4]. The exercise countermeasures proved effective, as evidenced by the fact that maximal oxygen uptake was maintained during flights that lasted up to 2 weeks [4]. Later in the Shuttle Program, the Extended Duration Orbiter Medical Project compared the effects on exercising and non-exercising crew. This project revealed crews that exercised with greatest intensity experienced the smallest reduction in maximal oxygen uptake and those crewmembers who exercised most frequently (more than three sessions per week) showed the smallest reduction in maximal oxygen uptake [4, 5].

As the ISS began operations, there was a greater emphasis on countermeasures due to the sheer length of the missions. The ISS is divided into the Russian Orbital Segment (ROS) and the United States Orbital Segment (USOS), which includes the ESA module (Columbus) and JAXA module (Japanese Experiment Module or JEM). We'll discuss the ROS countermeasure capabilities shortly but first the USOS side of the ISS. The primary exercise countermeasure devices in the USOS include the ARED, the FWED, CEVIS, and COLBERT (◘ Table 7.1). For a detailed description of the USOS countermeasure program, the reader is referred to Hackney and Loehr [6, 7].

Now, you may be wondering how effective all this exercise is. Well, that's difficult to say and here's why. While on board ISS, all long duration mission astronauts *must* abide by flight rules, which state quite clearly that all crewmembers must perform exercise. No exceptions. So, this means it is impossible to compare the effects of inflight exercise with no exercise. Having said that, there is a wealth of data from long-duration missions. For example, we know that the rate of BMD loss in ISS crew (3% at the lumbar spine and 6% at the hip) is less than measured in Mir crews who flew missions of between 117 and 438 days [10]. Having said that, there is a significant difference across crewmembers, with some astronauts losing up to 15% of their BMD following a 6-month mission. That equates to a disturbing 2.5% every month! [10]. Lately the trend is a gradual lowering of BMD loss in astronauts. One reason is the change in countermeasure equipment. In the early days of ISS, the go-to exercise equipment was the Interim Resistive Exercise Device (IRED). The IRED (◘ Fig. 7.7), which was designed by SpiraFlex Inc, provided astronauts with a linear resistance of up to 300 pounds and allowed crewmembers to perform squats to load their legs and spine which take the biggest hit in microgravity.

7

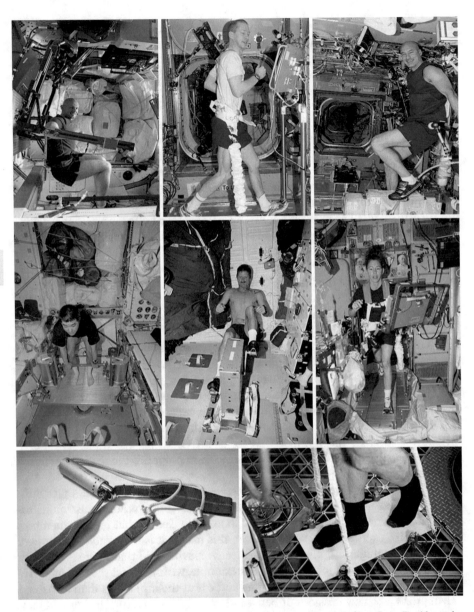

◘ **Fig. 7.5** Exercise countermeasures hardware over the years. Top row from L to R: Alexander Gerst exercises using the advanced resistive exercise device (ARED) on the ISS (Credit: ESA/NASA). Frank de Winne uses the T2 treadmill on ISS (Credit: NASA). Luca Parmitano uses the Cycle Ergometer with Vibration Isolation and Stabilization System (CEVIS) on ISS (Credit: ESA/NASA). Middle row L to R: Dan Tani, Expedition 16 Flight Engineer, uses the Interim Resistive Exercise Device (iRED) (Credit: NASA). Joe Tanner, STS-97 Mission Specialist, uses the cycle ergometer aboard Endeavour (Credit: NASA). Sandra Magnus, Expedition 18 Flight Engineer, uses the Treadmill with Vibration Isolation and Stabilization System (TVIS) (Credit: NASA). Bottom row L to R. The *Apollo Exerciser* used by Apollo 11 astronauts. (Credit Smithsonian National Air and Space Museum [NASM 2009-4775]). The Teflon-covered treadmill-like device used during Skylab 4 (Credit: NASA)

◘ **Table 7.1** Characteristics of the USOS Countermeasure Program [8, 9]

Consists of aerobic and resistance exercise

High-frequency program, comprising two sessions per day (1 x 30 to 45 minutes of aerobic exercise and 1 x 45 minutes of resistance exercise, 6 days per week

Multimodal, utilizing one resistance device such as the ARED and two aerobic devices such as the COLBERT (◘ Fig. 7.6) and the CEVIS

Aerobic and resistance sessions are completed on the same day, usually with a minimal break in-between because this saves time un-stowing/stowing exercise equipment

T2 allows running speeds up to 20.4 km/h with vertical loads equivalent to 54.4–68.0 kg

CEVIS provides workloads up to 350 W at 120 rpm

Aerobic sessions consist of steady state and interval-type protocols, with target intensities of 75–80 and 60–90% VO_{2max}

ARED engages all major muscle groups, with loads up to 272 kg

Resistance protocols are multi-set, multi-repetition for lower and upper body, with initial loads calculated from a 10-repetition maximum load (plus 75% of body weight to compensate for the absence of body weight) and adjusted thereafter based on performance

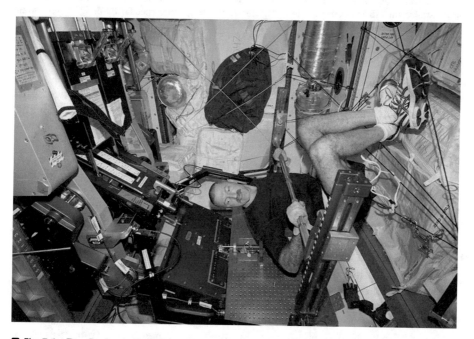

◘ **Fig. 7.6** Dan Burbank, Expedition 30 commander, works out using the ARED, which is housed in the Tranquility module of the ISS. Credit: NASA

7

◘ Fig. 7.7 JAXA astronaut Koichi Wakata, Expedition 20 Flight Engineer, completes his daily workout on the IRED, which at the time was located in ESA's Harmony module. The IRED, which was installed on ISS in 2000, was the first resistive exercise device specifically designed for use in space. It was used until 2009. Credit: NASA

Replacing IRED with ARED had a pronounced impact on BMD loss. In the days of IRED, astronauts could expect to lose between 3.7 and 6.6% across all bone sites, whereas in the days of ARED, astronauts may lose between 2.6 and 4.1%. Another bonus was the effect on muscle atrophy. Using ARED, astronauts nowadays lose between 8 and 17% less muscle across all sites [11].

Exercise Capacity

Of course, countermeasures for bone loss and muscle atrophy comprise only part of the suite of exercise equipment on ISS because astronauts must also maintain their exercise capacity. One metric for measuring aerobic capacity is maximal oxygen uptake, which measures the amount of oxygen utilized by each kilogram of body weight per minute of exercise. Astronauts perform preflight, inflight (◘ Fig. 7.8), and postflight measures of oxygen uptake, and, like all physical fitness metrics, there is an observed decline in crewmembers' ability to utilize oxygen. Why? First, there is the issue of muscle atrophy. Imagine losing 20 to 25 percent of your respiratory muscle mass (your intercostal and intracostal muscles). Obviously, that would make breathing more difficult. And then of course there is the loss of working (skeletal and cardiac) muscle mass. If your muscles – and heart! – are smaller, it stands to

Fig. 7.8 Dan Burbank performs a maximal oxygen uptake test while exercising on the Cycle Ergometer with Vibration Isolation and Stabilization (CEVIS). Credit: NASA

reason you will find it more difficult to work out. Compared with pre-flight data, maximal oxygen uptake declines by between 15 and 25 percent [1, 12].

Generally, the exercise countermeasure systems, the evolution of these systems, combined with the manipulation of exercise training regimes (continuous vs. interval, high intensity vs. low intensity), has enabled moderate levels of cardiovascular and cardiorespiratory adaptation. Having said that, as with bone loss, there is a significant amount of individual variation, so exercise countermeasures still must be optimized to take this variation into account.

With plans to venture further afield, optimizing exercise countermeasures will be a significant challenge. That is because the transport that agency astronauts (as opposed to certain obscenely wealthy commercial astronauts who may be venturing to the Red Planet on board the Starship) will be using to travel to Mars is NASA's Orion spacecraft, which has a habitable volume of less than nine cubic meters (compared with more than 900 cubic meters on ISS) – not enough room to swing the proverbial cat, never mind fitting in a variety of exercise countermeasures. So, replicating the exercise countermeasures and efficacy of ISS countermea-

Fig. 7.9 The interior of the Orion spacecraft. Credit: NASA

7

Table 7.2 Research on exercise countermeasures

Individual variation: real variation vs. within-subject random variation [13]

High-intensity interval training: efficacy and safety [14, 16]

Strength development and maintenance: the contribution of different training variables to the effectiveness of resistance training [15]

Concurrent training: the scheduling of aerobic and resistance exercise for maximizing training gains [17]

Plyometric/impact exercise: effects on the musculoskeletal *and* cardiovascular systems [11]

The role of nutrition in promoting adaptations to exercise training [18]

Complementary strategies: CM that could enhance the effects of or reduce reliance on exercise [19]

sures on exploration missions beyond Earth orbit will prove almost impossible. To that end, there has been a concerted effort to manipulate the exercise variables to try and develop more effective exercise countermeasures that *can* be performed in the confines of the Orion (■ Fig. 7.9). The variables that have been tweaked include mode, frequency, duration, workload, recovery, time under tension, and intensity. A snapshot of some of the research conducted in recent years is listed in ■ Table 7.2 below.

Radiation Countermeasures

In addition to helping astronauts maintain their fitness, another key function of any spacecraft life support system (LSS) is protecting astronauts from radiation.

Exactly how does radiation damage human physiology? An abundant molecule in us is water, a key component of which is oxygen. The presence of oxygen is important because molecular oxygen is key to the formation of reactive free radicals. This means that in areas of high concentrations of oxygen, the effects of radiation are increased. Conversely, in areas of low oxygen concentration, tissues and cells are protected thanks to hypoxia.

Of all the free radicals, the *hydroxyl* radical is one of the most damaging. One of the structures that this free radical damages is DNA, a key structure for cell survival [20]. The hydroxyl radical, together with other oxidative radicals, is responsible for double-strand breaks (DSBs) and base lesions. Although the body goes to work repairing the damage, the process is not always successful, since if two base lesions on opposite strands are too close, a double-strand break may occur.

This mechanism is known as the DNA damage response (DDR), and it is affected by the radiation dose, the type of radiation, and the amount of tissue exposed to that radiation. The damaged DNA material is repaired thanks to the action of key proteins (MRE_{11}, $RAD_{50,}$ and NBS_1) which determine the damage before binding the broken DNA strands together in a process that can proceed along two pathways. One of these pathways is homology-directed repair (HDR) which is a process that results in no errors. The other pathway is non-homologous end joining (NHEJ), which is a process that results in sequence deletions. If cells are not repaired successfully, genomic instability can result, which in turn can cause mutations and carcinogenesis. The extent of the mutations and the degree of risk of carcinogenesis will depend on the type of tissue, the radiation dose, and the amount of tissue exposed to radiation [20]. How these mutations may occur is not fully understood, but it is probably a result of damage to progenitor cells, blood vessels, and persistent oxidative stress. An extra risk is repeated exposure to radiation, which is definitely something deep space astronauts will be exposed to. In this case, repeated inflammation will be the result, which will lead to fibrosis and increased DNA damage, with possible outcomes being death, or at the very least, permanent tissue damage.

Setting Acceptable Risk Levels for Astronauts

The permissible exposure levels (PELs) for radiation that NASA sets for its astronauts are set to prevent inflight risks and to limit risk to a level that is acceptable from an ethical, moral, and financial standpoint. In the 1960s and 1970s, PELs were set based on recommendations made by the National Academy of Sciences (NAS). In the 1980s, more data on radiation exposure had been accumulated, and NASA asked the National Council on Radiation Protection (NCRP) to reassess the dose limits for astronaut working in low Earth orbit. This reassessment culminated in the NCRP Report No. 98, published in 1989, which recommended age and gender career dose limits that applied a 3 percent increase in cancer mortality as a risk limit.

NCRP Report No. 98 was followed by revisions to the acceptable level of radiation risk in LEO in NCRP reports published in 1997 and 2000 [21]. The astronaut

◘ Table 7.3   Theoretical dose limits for 1-year missions based on 3% REID[a]

Age at exposure	E (mSv) for 3% REID (average life-loss per death [y])	
	Males	**Females**
30	620 (15.7)	470 (15.7)
35	720 (15.4)	550 (15.3)
40	800 (15.0)	620 (14.7)
45	950 (14.2)	750 (14.0)
50	1150 (12.5)	920 (13.2)
55	1470 (11.5)	1120 (12.2)

[a]Adapted from: Radiation Risk acceptability and limitations. Cucinotta F. (▶ https://three.jsc. nasa.gov/articles/AstronautRadLimitsFC.pdf). 12-21-2010

risk level of a 3 percent[1] increase risk in cancer over a lifetime is similar to the risk level applied to workers in nuclear facilities. The difference is that the radiation doses that nuclear workers are exposed to correspond to a lifetime of exposure to relatively low radiation doses compared with the exposure limit in LEO. For example, radiation workers in nuclear reactors rarely approach an annual average exposure of 2 mSv. This is significantly below the effective dose of 80 mSv that astronauts are exposed to during a 6-month increment on board the ISS. The point is that ground workers are exposed to *chronic* exposure at low levels (compared with radiation levels in LEO) of radiation over a long period of time, whereas astronauts are exposed both chronic *and* acute radiation.

Permissible Exposure Limits

NASA's PELs take into account a number of factors, including age, gender, latency effects, differences in tissue types, and differences in lifespan between genders [22]. When all these factors are considered, a risk projection calculation can be made, and a risk of exposure-induced death (REID) from fatal cancer can be made (◘ Table 7.3).

Another limit that NASA applies is one for non-cancer effects. For example, radiation exposure may also cause prodromal effects such as nausea and fatigue, and it may also cause heart disease, dementia, and central nervous system (CNS)

---

1   There are many approaches to setting acceptable risk levels. One is to set an unlimited risk level, but this would not be popular with astronauts or their families. Another option is to base risk on life loss from radiation-induced cancer against cancer deaths in the general population. The current method uses the reference point of a ground-based radiation worker.

■ **Table 7.4a** Dose limits for short-term and career non-cancer effects[a]

Organ	30-day limit	1-year limit	Career limit
Lens	1000 mGy-Eq[b]	2000 mGy-Eq	4000 mGy-Eq
Skin	1500	3000	6000
Blood-forming organs	250	500	N/A
Heart	250	500	1000
Central nervous system	500	1000	1500

[a]Adapted from: Radiation Health Risk Projections Briefing to NAC HEOMD/SMD Joint Committee 7 April 2015
[b]Milli-Gray Equivalent. The Standard International (SI) unit of absorbed dose is the gray (Gy). 1 Gy is a measure of the absorption of one joule of radiation energy per kilogram. Note: the gray is different from the Sievert (Sv), which is the SI unit that represents the biological effect of radiation

■ **Table 7.4b** Tissue weighting factor calculated by attributing an estimate for a tissue's contribution to cancer

Organ	Tissue weighting factor	Organ	Tissue weighting factor
Gonads	0.20	Liver	0.05
Bone marrow (red)	0.12	Esophagus	0.05
Colon	0.12	Thyroid	0.05
Lung	0.12	Skin	0.01
Stomach	0.12	Bone Surface	0.01
Bladder	0.05	Remainder	0.05
Breast	0.05	Adrenals, brain, intestine, kidney, muscle, spleen	

damage. For dose limits for non-cancer effects, NASA calculates the relative biological effectiveness (RBE) factor to the major organs of the body as indicated in ■ Tables 7.4a and b.

Other space agencies such as the European Space Agency (ESA) and the Russian Space Agency (RSA) estimate dose limits based on data published by the International Commission on Radiological Protection (ICRP) [23]. While these dose limits can be applied to LEO workers, there are no operationally approved dose limits for deep space, although estimates have been made for the risk astronauts will be subjected to when traveling to the Moon (■ Fig. 7.10) or Mars (■ Table 7.5).

Fig. 7.10 The Artemis Program will return astronauts to the Moon by 2024. Credit: NASA

Table 7.5 Calculation of %-REID from fatal cancer for lunar or Mars missions at solar minimum behind a 5-g/cm^2 aluminum shield.[a] The effective dose, E, is averaged over tissues susceptible to cancer risk

Mission type	E, Sv	%-REID
	Males (40 years old)	
Lunar (180 days)	0.17	0.68
Mars swingby (600 days)	1.03	4.0
Mars exploration (1000 days)	1.07	4.2
	Females (40 years old)	
Lunar (180 days)	0.17	0.82
Mars swingby (600 days)	1.03	4.9
Mars exploration (1000 days)	1.07	5.1

[a]Adapted from Cucinotta and Durante (2006)

The numbers presented in ◘ Table 7.5 are not encouraging for space agencies hoping to send astronauts to the Moon for lengthy stays or to Mars (◘ Fig. 7.11). A lunar stay of 180 days results in an exposure of 170 mSv, which is more than twice that of a 180-day stay on board the ISS. But a roundtrip to Mars results in an exposure of more than 1000 mSv, which exceeds NASA's guidelines (◘ Tables 7.6a and b) stating that no astronaut be exposed to more

Fig. 7.11 NASA's Journey to Mars. Credit: NASA

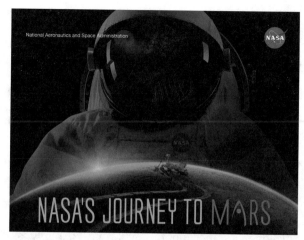

Table 7.6a Career exposure limits for NASA Astronauts[a]

Age	25	35	45	55
Male	1.50 Sv	2.50 Sv	3.25 Sv	4.00 Sv
Females	1.00 Sv	1.75 Sv	2.50 Sv	3.00 Sv

[a]An astronaut's organ and career exposure limits are determined by age and gender. An average does for an Earthbound person is 0.0036 Sv, whereas someone who works in a nuclear power plant may be exposed to as much as 0.05 Sv per year without exceeding International Standards. As you can see, the limit for astronauts is significantly higher

Table 7.6b Depth of radiation penetration and exposure limits for astronauts and public (Sv)

	Exposure interval	Blood-forming organs (5 cm depth)	Eyes (0.3 cm depth)	Skin (0.01 cm depth)
Astronauts	30 days	0.25	1.0	1.5
	Annual	0.50	2.0	3.0
	Career	1–4	4.0	6.0
General public	Annual	0.001	0.015	0.05

than this amount of radiation in a lifetime. To exceed this limit is to increase the risk of developing a fatal cancer by 5 percent or more. For astronauts embarking on such a trip, the accumulated radiation dose would be akin to getting a whole body computed tomography (CT) scan every 5 days [24, 25]. Also, exposure to more than 1,000 mSv would not amount to a "measured dose" (i.e., over a lifetime) because crewmembers would receive this amount in less than 3 years. Being exposed to such a large amount of radiation in such a short period of time would result in changes at the cellular level and possibly mild acute radiation syndrome (ARS) symptoms.

The ALARA Principle

While NASA may be able to reduce the amount of radiation exposure by extra shielding, another option might be to apply a different risk strategy [26]. Traditionally, the agency has applied the maxim *As Low As Reasonably Achievable* (ALARA) when it comes to exposing astronauts to radiation. Another option may be to change this principle to *As High As Relatively Safe* (AHARS). Since astronauts are exposed to so much radiation, they are classified as radiation workers. But because they are exposed to so much radiation, the amount of radiation that astronauts are exposed to exceeds all terrestrial limits. Which is why the Occupational Safety and Health Administration (OSHA) gave NASA a waiver that allowed the agency to create its own guidelines as outlined in this chapter. The radiation risk may be reduced with the operationalization of faster propulsion systems such as VASIMR (◘ Fig. 7.12), or it may be reduced with better shielding. But until either or both of these technologies becomes available, radiation risk will remain an intractable problem.

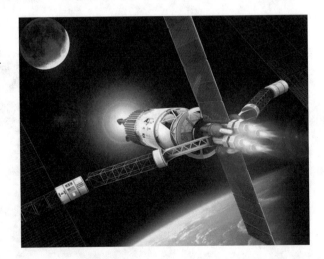

◘ **Fig. 7.12** Speed is everything when it comes to keeping astronauts safe. And VASIMR (Variable Specific Impulse Magnetoplasma Rocket) fits the bill. Capable of speeds approaching 230,000 kmh, this spacecraft could reach Mars in less than 6 weeks. Credit: Ad Astra

Radiation Dosimetry and Detection

Radiation monitoring on board the ISS is conducted to gather, analyze, and characterize the radiation environment to better ensure crew health [26, 27]. And thanks to careful consideration of radiation effects, the station does a very good job of protecting the crew. But given that astronauts are exposed to about 80 times the terrestrial radiation dose, radiation exposure remains a limiting effect on an astronaut's career, which is why it is important to accurately monitor exposure as closely as possible. To protect ISS crews from the effects of radiation, space agencies are guided by the *ISS Medical Operations Requirement Document* (MORD). The MORD identifies radiation exposure monitoring requirements as follows:

i.   Radiation doses absorbed by human tissue
ii. Charged particles and neutron radiation inside the ISS
iii. Charged particles outside the ISS during spacewalks

To quantify the station's internal and external radiation, NASA[2] has installed various *active* and *passive* radiation instruments that measure and document each astronauts' radiation exposure. In addition to the passive and active radiation instruments, each astronaut is provided with a dosimeter that serves as the dosimeter of record [26, 27]. The data from each astronauts' dosimeter, when combined with data from internal and external dosimeters, provides an accurate characterization of the radiation environment. This data can then be applied to exposure limits. Terrestrial exposure limits are much too restrictive if applied to the space environment. So space agencies have adopted recommendations made by the National Council on Radiation Protection (NCRP) that sets a career limit at a three percent REID from cancer.

But how does this number compare with the incidence of cancer in the general population? Here on Earth, the risk of developing and dying from cancer is about 20 percent, which means about 20 people out of 100 will die from cancer. But if you happen to be an astronaut, then that risk increases by three percent, which means 23 astronauts out of every 100 astronauts may die [26]. Of course, this metric is skewed because astronauts are perhaps the healthiest individuals on and off the planet.

Passive Dosimetry

One way of measuring and tracking radiation is to use passive dosimeters. These are placed at fixed locations inside the pressurized modules. The information from these dosimeters provide the ground with information about those locations where

---

2   Career radiation exposures of astronauts are tracked by a NASA radiation specialist who logs radiation exposures for each astronaut in a document known as the Astronaut Annual Radiation Exposure Report.

the exposure rate is high. This information can then be used to re-evaluate the amount of time each astronaut spends in each module.

Passive radiation dosimeters (PRDs) are also known as radiation area monitors (RAMs). Each RAM comprises a set of thermoluminescent detectors (TLDs) surrounded by Lexan. The material reacts to radiation by means of excitation of materials that comprise the TLDs. TLDs comprise lithium fluoride or calcium fluoride embedded in a solid crystal structure. When the TLD is exposed to radiation, the radiation interacts with the crystal. Some atoms in the crystal absorb the energy and become ionized. This generates free electrons and heats the crystal, which causes the material to vibrate, which in turn releases electrons. When electrons return to their original pre-ionized state, they release stored energy that appears as light. This light is measured using special photomultiplier tubes, and the amount of light – the "glow curve" – released is proportional to the amount of radiation that struck the crystal [26]. Crew passive dosimeters (CPDs) are provided to each astronaut. Apart from sporting a different label, the CPDs are exactly the same as the RAMs. They are worn by the astronauts throughout the mission, including spacewalks.

Active radiation monitors provide data to the ground that is used together with data provided by the CPDs to generate high-dose rate and low-dose rate areas inside the station [26]. These measurements also help the ground to reduce uncertainty when calculating risk assessments for the crew, which is done by evaluating the following metrics:

1. Linear energy transfer spectra inside the ISS
2. Mass distribution
3. Space weather conditions
4. Stage of the solar cycle
5. CPD data
6. RAM data

The tissue equivalent proportional counter (TEPC) uses gas to measure the dose of radiation. The function of the TEPC is to develop an exposure history of the crew during their stay on the ISS. It collects data by making spectral measurements of the energy loss of radiation as the radiation passes through a detector volume. Inside the TEPC is an omnidirectional detector encased in tissue equivalent plastic, similar to that used in the Matroshka Human Phantom [28, 29], discussed later in this chapter. Also inside the TEPC is propane gas that, combined with the plastic, provides an energy deposition effect similar to human tissue. This energy deposition response is achieved thanks to propane gas, which is kept at very low pressure. This means that radiation passing through the detector gas passes through with similar linear energy loss as a human cell. This information is stored inside a 512-channel spectrometer that is presented to the crew via an electronic display that displays total dose, total dose equivalent, and incremental dose. The TEPC also downlinks information to the ground for analysis. The TEPC became operational in 2000 and as of 2020 is still functioning (there is one active and one spare unit on board the ISS).

The charged particle directional spectrometer (CPDS) monitors the internal radiation environment of the ISS using a Cherenkov detector. The Cherenkov detector works by measuring Cherenkov light. The principle by which the detector works is as follows. A particle that passes through material (in this case a 12-element silicon stack) at a speed faster than the speed at which light can pass through the material emits light. An analogy is the sonic boom generated by an aircraft moving through the air faster than the sound waves can move through the air. In the case of the Cherenkov detector, the amount of radiation can be calculated if the angle *and* direction of light are known.

Intravehicular TEPC

The intravehicular TEPC (IV-TEPC) device measures radiation in near real time [26]. If the level of radiation exceeds predetermined thresholds, the device signals a caution and warning (C&W) alarm. In such an event, the crew would probably relocate to a higher shielded module. The unit itself is portable and comprises several tissue equivalent radiation detectors constructed of material and gas that react to radiation in a similar way that human tissue does. The IV-TEPC provides the following capabilities [26]:

1. Signal conditioning
2. Data manipulation
3. Storage
4. Real-time telemetry
5. Extended data download

The IV-TEPC was declared operational on board the ISS in March 2001. It failed in 2006 and has not been replaced.

European Crew Personal Active Dosimeter

The European Crew Personal Active Dosimeter (Escaped) is a device that assesses radiation exposure of European astronauts. The device comprises three main elements:

1. Mobile Unit
   - Two silicon detector modules
   - Absorbed dose detector
2. Personal Storage Device
   - TEPC
   - Internal mobile unit
   - Storage and charging capability for mobile unit
   - Local data analysis
   - Display of radiation data
3. Ground Station Analysis Software
   - Calculates the dose equivalent based on spectroscopic data

The mobile units (MUs) are battery driven and have a power and data interface with Columbus, where astronauts can access radiation data and where European Crew Active Dosimetry Activity is monitored in conjunction with the ground.

PADLES

The Passive Dosimeter for Lifescience Experiments in Space (PADLES) is used to monitor radiation inside the Japanese Experiment Module (*Kibo*). *Bio* PADLES is used to measure the biological effects of radiation, while *Area* PADLES is used to assess the personal exposure of each astronaut, and *Free-Space* PADLES measures the radiation environment outside Kibo. The key radiation-sensitive material inside the detectors is CR-39 plastic, and the radiation measured by the dosimeters is published in JAXA's PADLES database.[3] Operationally, *Free-Space* PADLES is launched and returned via pressurized cargo on board the Cygnus or Dragon, packed inside a Crew Transfer Bag. On station, *Free-Space* PADLES is installed on the Multi-Purpose Experiment Platform inside the JEM. It is then moved outside the JEM via the JEM airlock and is grabbed by the JEM RMS robot arm which positions *Free-Space* PADLES on the exterior of the module.

Detectors

The ISS flies at an altitude of around 400 kilometers. At this altitude, crews are provided with a safe space platform that is protected against most of the deleterious effects of ionizing radiation. Over the years of ISS operation, the best *active* dosimeters have proven to be those that provide linear energy transfer (LET) data and tissue equivalent proportional data (such as the TEPCs) [30]. Supporting the data provided by the dosimeters is a suite of detectors. Together, these dosimeters and detectors are used to provide a very accurate set of baseline data that allow scientists to benchmark risk assessments for long-duration flights, as long as those long-duration flights occur in LEO. Beyond LEO? Well, that's another story (◘ Table 7.7).

Matroshka

Matroshka is an experiment sponsored by Roscosmos, ESA, and JAXA. It comprises a human phantom that is used to measure radiation astronauts are exposed to inside and outside the ISS [28, 36]. The phantom is designed using human tissue equivalent material and is filled with water and instrumented with a series of passive radiation detectors (◘ Fig. 7.13).

Between 2004 and 2009, Matroshka was exposed to radiation on three occasions – two inside the ROS and one outside the ISS. After crunching the numbers, scientists at the German Space Agency and the Technical University in Vienna revealed that individual dosimeters worn by astronauts inside ISS had overestimated radiation exposure by 15% compared with the actual dose measured inside Matroshka. The overestimation in space was more than 200%.

---

3   ▶ http://idb.exst.jaxa.jp/db_data/padles/NI005.html.

◘ **Table 7.7**   A Selection of the detectors used on board the ISS

Liulin. The Liulin system utilizes silicon detectors to measure deposited radiation energy [31, 32]. Basically, the number of charged particles that hit the device converts to a dose rate. The first version (Liulin-E094) was first used on board the ISS in April 2001 and was followed by a series of updated systems that were flown inside the ISS (Liulin ISS, between September 2005 and June 2014, and Liulin-5, which was deployed between May 2007 to present) and outside the ISS (R3DE, from February 2008 to September 2009 and R3DR between March 2009 and August 2010)

Alteino. This detector is also referred to as SilEye3. It made its first appearance on the *Mir* space station. The shoebox-sized system comprises a stack of eight silicon striped sensors measuring 80 mm by 80 mm by 0.38 mm and two plastic scintillators [33, 34]. The orientation of the stack is configured to provide a set of three coordinates through which particles strike. This configuration allows for tracking the direction of particles

ALTEA. This is a system of six silicon telescopes similar to Alteino except for the scintillators. These detectors have been positioned in the US Lab in three axes and also in Columbus [34, 35]. The system downlinks data via real-time telemetry

DosTel. This detector system is also based on silicon detectors. They are deployed in Columbus [27]

TRITEL. This system comprises a set of three silicon telescopes configured in a three-dimensional arrangement. The system has been deployed in Columbus (in the European Physiology Module rack TRITEL-SURE) since 2012

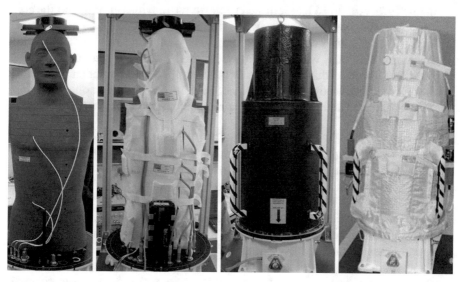

◘ **Fig. 7.13**   Matroshka. The phantom is essentially a torso comprising 33 slices each of 2.5-centimeter thickness. Each slice houses thermoluminescent lithium fluoride detectors (about 4.5 millimeters in diameter) placed in plastic tubes. Thanks to the positioning of the TLDs inside the phantom, it is possible for scientists to accurately determine the spatial distribution of radiation and thereby calculate the effective dose. The key to the way radiation is measured is found inside the TLDs. Inside each detector is a lattice, which traps free electrons created by radiation. The greater the radiation dose, the greater the number of trapped electrons [36]. When exposed to heat, the trapped electrons are released, emitting light, and it is this light that provides the index for radiation exposure: the greater the light, the higher the proportional radiation dose. Credit: ESA

» We must remember that measurements within the MATROSHKA experiment were performed at low Earth orbit where the Earth's magnetosphere significantly reduces the number of charged particles from cosmic radiation. In interplanetary space there is no such shielding.

Dr. Bilski, a MATROSKA scientist suggesting that manned Mars missions may still be a risky proposition despite the lower than expected radiation levels measured on ISS

Shielding

» The space radiation environment will be a critical consideration for everything in the astronauts' daily lives, both on the journeys between Earth and Mars and on the surface. You're constantly being bombarded by some amount of radiation.

Ruthan Lewis, architect and engineer at NASA's Goddard Space Flight Center

7 August 1972. An enormous flare exploded from the Sun spewing out a burst of energetic particles. A moonwalker caught in the storm would have been exposed to 400 rem. Not necessarily deadly but enough to require a mission abort and an early return to Earth. Fortunately, there were no astronauts on the surface of the Moon in August 1972. Apollo 16 had returned to Earth the previous month, and the crew of Apollo 17 was preparing for a mission that was due to take place in December that year. Of course, an astronaut isn't going to be wandering around on the Moon when a storm hits. They will be ensconced inside their base or spacecraft. If such an event had occurred during an Apollo mission, the Apollo command module's hull[4] would have reduced the 400 rem to about 35 to 40 rem. Still enough of a dose to cause a headache and perhaps some nausea, but not sufficient to require a bone marrow transplant.

» There's a lot of good science to be done on Mars, but a trip to interplanetary space carries more radiation risk than working in low-Earth orbit.

Jonathan Pellish, space radiation engineer at Goddard

In science fiction movies, the most dangerous threat to the crew is usually some form of alien life. And in most such movies, these threats are usually pretty big. But in the real world of sending astronauts on deep space interplanetary missions (■ Fig. 7.14), the dangers are mostly invisible. Heavy elementary particles zipping along through space and tearing through DNA are enough to give any astronaut serious concern. These cosmic rays present irreducible risks and are as deadly as any threat Hollywood can conjure up. So, how do we protect astronauts from this danger?

---

4   The Apollo Command Module's hull provided 8 g/cm² or radiation protection. The Space Shuttle had 11 g/cm², and the ISS has up to 15 g/cm² at its most shielded areas. In contrast, a spacesuit has 0.25 g/cm².

◘ Fig. 7.14 One of the biggest dangers to future manned interplanetary missions will be radiation exposure. Credit: NASA

Here on Earth, the sheer bulk of the atmosphere does a great job shielding us from the worst shrapnel-like cosmic rays can inflict. Many miles above us, incoming protons are absorbed by the nuclei of air atoms. Particles and subparticles disperse in a series of annihilating cycles until all that is left are some peons and mesons, some of which pass through our body. But at that stage, there is so little energy left in them – thanks to the weight of the atmosphere – that all they can do is produce a few ions.

Above the atmosphere and beyond LEO, the situation is very different. In deep space, those same cosmic rays have nothing to disperse them. Except spacecraft. And the astronauts inside them. And once those heavy nuclei zip through the skin of the spacecraft and through the human body, the trail of damage is devastating. Broken bonds. Genetic material ripped apart. Tissues permanently damaged. The body has an extraordinary capacity for self-repair, so a week or month of this is survivable. But two or more years? Unlikely. We know this from studying the grave biological consequences of the unfortunate humans who have been exposed to intense bursts of radiation.

The Mars evangelists promote the fact that since some astronauts have spent 6 months in space, traveling to Mars should be a breeze. But astronauts on board the ISS are still shielded by the Earth's magnetic field.

What are the solutions? Shielding perhaps? One shield suggested by those in the business of protecting astronauts during exploration class missions (ECMs) is a sphere of water. The only drawback is that such a system would weigh 400 tons at minimum! How about a superconducting magnet? Such a system would use a magnetic field to repel cosmic rays, but the problem is that the magnetic field itself would present certain health risks.

So what other shielding solutions are out there? Before we discuss these, it's important to have a reference for what exactly engineers are up against. We'll begin with the legal limit for those working in nuclear power plants. That number happens to be 2 mSv per year. A Mars astronaut by comparison would be exposed to 1000 mSv per year, and the consequences of that exposure would be that one in ten male interplanetary astronauts would die from cancer. Many more would suffer from radiation-induced cataracts and brain damage [22]. And that's a best case scenario, because it isn't just cosmic rays that inflict damage. Every once in a while, the Sun unleashes huge bursts of heavy nuclei that travel at close to light speed. These bursts, which can deliver more than 100 mSv per hour are, quite simply, a death sentence for any unshielded astronaut in deep space.

7   Water as a Radiation Shield

So what about those options mentioned earlier? Before considering the use of water (astronauts need this, so it makes sense to use it) as a shielding material, we need to perform some basic calculations. First, we need to know how much shielding material it takes to protect an astronaut. If we wanted to provide an interplanetary astronaut with the same shielding as on Earth, it would take one kilogram of water per square centimeter. But, since astronauts are willing to accept risk, let's give them less – and more affordable – protection and have them make do with just 500 grams of water per square centimeter. That amount of shielding is equivalent to you living at an altitude of 5500 meters. Now for the sake of simplicity, let's make our spacecraft a sphere. To protect our crew with water, the walls of this spherical vehicle would need to be five meters thick and would weigh about 500 tons. Back in the old days, the Space Shuttle could ferry around 30 tons into LEO. The Space Launch System? 130 tons in its most powerful lift configuration. So, 500 tons is too heavy. But what if the engineers reduced the amount of water and increased the *hydrogen* content of the spacecraft walls? They could do this by using polyethylene and perhaps bring the weight down to 400 tons. That just isn't financially feasible, so let's consider the other option mentioned – magnetic shielding.

Magnetic Shielding

This is yet another exotic shielding option promoted by the Mars evangelists. Problem solved! Well, no, the problem is not solved. That is because this method of shielding hasn't moved far beyond the PowerPoint phase of development, and there are good reasons for this.

Let me explain. Earth is surrounded by a magnetic field that does a great job deflecting incoming charged particles so it would seem reasonable to assume – unless you happen to be a particle physicist – a spacecraft could carry a magnet to do the same. The problem is those cosmic rays. These have tremendous kinetic energy, and trying to bring them to a standstill in the space of just a couple of meters requires energy the likes of which might be achievable in the world of Star Trek but in the real world is not. It would require a magnetic field of 20 tesla to stop cosmic rays, and 20 tesla is about 600,000 times the strength of the Earth's magnetic field. How would humans endure living in a magnetic field of 20 tesla and what would the long-term effects be? Volunteers form a line here, please!

But, some insist, it might be possible to use a second magnet to cancel out the field effect of the first magnet. Such a system, its proponents argue, could use plasma to push away the magnetic field of the first. But the problem with plasma is that is very unstable, and even if it could be controlled, the nuances of how plasma behaves in a magnetic field would mean the field would be weakened, not strengthened.

Electrical Shielding

So, water is too expensive, and the magnetic option is too tricky and downright dangerous. What about electrical fields? In this application, the spacecraft would be charged electrically with *two billion volts*! Such a charge would, in theory, repel cosmic ray protons. In theory, the problem is that space – even deep space – is not empty. Even in deep space, there are ions and electrons flying around, and these negatively charged electrons would be attracted by a spacecraft that is positively charged. Let's not forget that this spacecraft would have an electric field that would extend tens of thousands of kilometers away from the vehicle. Such a huge electric field would draw in electrons from a huge volume of space, and when those electrons hit the walls of the spacecraft, they would act just like the cosmic rays the shield was designed to repel! The electrons would generate gamma rays as soon as they hit the vehicle, and the intensity of this barrage would be so great that it would put the original headache in the shade. And what about those two billion volts? Does anyone have any idea of what sort of system could generate such a current? Two billion volts is 2000 megawatts, which is about the same amount of power generated by your average power plant. How do you fit such a system on an interplanetary spacecraft? The answers to these questions are few and far between.

Linear Energy Transfer and Relative Biological Effectiveness

One process that is key to understanding which materials make the best shields is the process of how radiation interacts with the spacecraft and the occupants inside. That is because radiation does not simply pass through the walls of a spacecraft, just like it does not simply pass through the bodies of the astronauts. Radiation *interacts*. And as it interacts, the energy of all that ionizing radiation is disrupted and the size of the particles reduced [37, 38, 39]. The problem is that the disruption causes the heavy charged particles – *primary radiation* – to disintegrate into smaller particles, *secondary radiation*, and it is these smaller particles that cause biological damage in the crew.

Now you might think the solution would be to conduct research that mimics this interactive process, but that is not what research does. Instead, almost all research studies that simulate the effect of galactic cosmic rays (GCRs) do so by exposing animals to heavy-ion accelerators to simply replicate the dose a human crew might be exposed to during an interplanetary mission. This method does not provide a true model of what happens in deep space, because it is very difficult to replicate the myriad energies of disrupted GCRs and even more difficult to measure the extent to which these energies inhibit cell regrowth and tissue repair mechanisms. Furthermore, different animals respond differently to radiation. Some are more susceptible and some less sensitive to radiation damage. And finally, the current technology of heavy-ion accelerators is limiting in terms of accurately reproducing the ions in the GCR spectrum.

So what other metrics can be applied to simulate the effects of GCRs? One way researchers simulate the effects of radiation is to apply the metric of linear energy transfer or LET. LET is used to measure the amount of tissue damage caused by radiation, and it is a metric used to determine radiation protection and risk assessment. This metric is used in conjunction with the measure of relative biological effectiveness (RBE), which is a measure applied to the effect of different types of radiation. In essence, the higher the RBE for a specific type of radiation, the more damaging that radiation is per unit of energy when it is absorbed by human tissues. Several studies utilizing LET and RBE have been conducted over the years by measuring the LET spectrums using TEPCs, and plastic nuclear track detectors placed at different locations on board the ISS. Similar studies were conducted during NASA's Exploration Flight Test (EFT-1), which tested the Orion Multi-Purpose Crew Vehicle (MPCV) during orbital flight. While the 4-hour EFT-1 flight was much

Spallation

When high-charged particles penetrate shielding or astronauts, they do so in a straight path to begin with. But shortly after penetrating matter, those heavy ions begin to disperse as they collide with atoms in the shielding and/or the astronauts. As the path of the heavy ions is disrupted, energy is dissipated, but at the same time, smaller nuclei are generated in a process called *spallation*. The degree to which energy is dissipated is largely determined by the properties of the material through which the heavy particle are traveling. Generally, energy loss increases with decreasing atomic number, which is why hydrogen is such an effective shielding material. Scientists can calculate the stopping power of a material by determining the energy lost per unit path length that the particles travel. This number is the LET, which is a metric that quantifies how much energy is lost as the heavy ions transit material. But stopping power isn't everything. A good shielding material should not only stop as many of the high-energy particles as possible, but also limit the amount of fragmentation as much as possible, in addition to being able to stop as many of the low-energy particles as possible. Polymers tend to be good candidate shield materials because they have a high hydrogen content and also stop more low-energy particles than most other materials. But the choice of the material is just one consideration. The next decision is deciding how thick the material should be. This is important because the LET of the heaviest nuclei has such penetrating energy that they can travel deep through a material before there is any measurable energy loss. It is therefore important that the shielding material is designed in such a way that spallation is limited and energy loss is maximized. This is very difficult to do because the data on fluence of particles in deep space is limited (one application that is used to calculate this is the Monte Carlo particle transport simulation software PHITS). Even with advanced simulation software such as PHITS, reliable and accurate predictions of how well a shield will function are very difficult. This is because of the lack of data from deep space and the difficulty in predicting how neutron propagation (which is highly sporadic) occurs in biological tissue.

shorter than 6-month ISS flights, the high apogee (5800 kilometers) of the second orbit included a transit through the radiation-dense Van Allen belts and also a brief excursion into the interplanetary environment. Radiation detectors were activated shortly after liftoff and collected radiation data for the duration of the flight.

Polyethylene as a Shielding Material

One candidate for radiation shielding is polyethylene, a plastic that is also found in water bottles. By virtue of its high hydrogen content [40] and the fact the material is very cheap, this material also offers other advantages when it comes to protecting astronauts from radiation. For example, plastic-like materials such as polyethylene cause much less secondary radiation than traditional materials such as aluminum. Since polyethylene isn't the most versatile material when it comes to building spacecraft, the material has been adapted to a stronger and lighter material: RXF1. Created by Raj Kaul, RXF1 is derived from polyethylene, but since it is a fabric, it can be shaped into whatever shape is needed. While polyethylene has been shown to be effective at dispersing heavy ions, stopping protons, and slowing down neutrons (which form as secondary radiation), it is not a structural material, although RXF1 may have some potential in that application. But there is another hydrogen-based material that might do both jobs. Hydrogenated boron nitride nanotubes (❏ Fig. 7.15), also known as hydrogenated BNNTs [40, 41], comprise nanotubes constructed of carbon, boron, nitrogen, and hydrogen. In addition to absorbing secondary neutrons and stopping

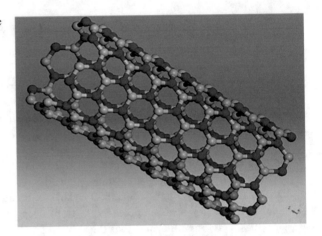

◘ **Fig. 7.15**   BNNTs may prove to be an effective shield against radiation. Credit: NASA

protons, the hydrogenated BNNT material is so flexible that spacesuits could be made of it.

>> This product will enable human deep space exploration. Our breakthrough has come in creating the architecture of the multi-layered shield to accurately cover the most important organs.
>>   Oren Milstein, CEO, StemRad

AstroRad Radiation Shield

Another passive means of protecting astronauts is a vest being developed by Israeli researchers. Dubbed the AstroRad Radiation Shield, the vest is being produced by StemRad, a company based in Tel Aviv. The vest (◘ Fig. 7.16), which will be customized for each crewmember, is designed to protect vital organs. To test the concept, the vest will be "worn" by a phantom torso that will measure radiation absorption. Another phantom torso will be flown that will be unprotected.

Pharmacological Countermeasures and Radioprotectors

Space agencies conduct radiation research because astronauts are exposed to *chronic* doses of radiation. But during long-duration missions beyond LEO, there is a real danger that crews may be exposed to *acute* doses that may lead to *acute radiation syndrome* (ARS). To be prepared for such missions, space agencies must be prepared to anticipate radiation exposures and be able to deal with the consequences. One way to do this is to implement a radiation medical coun-

◘ Fig. 7.16 Astrorad: Perhaps one of the more elegant ways to shield astronauts against radiation. Credit: Stemrad/NASA

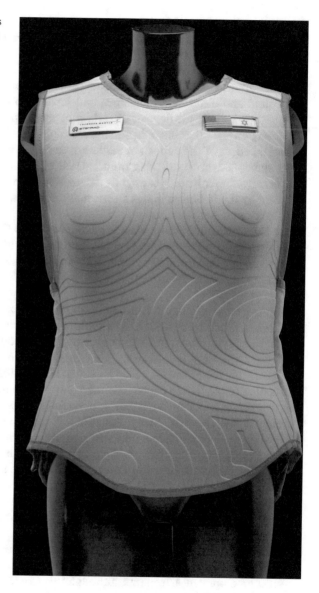

termeasures program that would cover products to be used following a radiological emergency.

Radioprotectors are compounds that can be considered as a preemptive medical countermeasure, since they protect against radiation injury and the effects of ionizing radiation only when administered *before* any radiation exposure. This is different than a *mitigator*, which protects against radiation injury *after* exposure to radiation. Research that studies radioprotectors and mitigators usually investigate the effects of acute total body irradiation (TBI) in rats. While TBI affects several organ systems, death in the first 30 days, whether in rats or humans, is usually the result of two mechanisms:

1. Gastrointestinal Syndrome

    Death within 10 to 12 days after exposure of 8 to 20 Gray, usually as a result of fluid and electrolyte imbalance and sepsis. Note a *Gray* (Gy) is a physical quantity. 1 Gy is the deposit of radiation energy per kg of matter or tissue. A Sievert on the other hand represents a biological effect, i.e., the equivalent biological effect of the deposit of radiation energy in a kilogram of human tissue. In someone who is suffering from 1. Gastrointestinal Syndrome, the fluid and electrolyte imbalance is caused by a depletion of intestinal stem cells, which are killed by the radiation in a prs known as apoptosis.

2. Hematopoietic Syndrome

    Death within 30 days after exposure to 3 to 8 Gray, usually as a result of neutropenia and thrombocytopenia. In someone suffering from this syndrome, neutropenia and thrombocytopenia is caused by the depletion of radiosensitive hematopoietic progenitor cells for white blood cells.

To improve survival rates in astronauts who may be exposed to very high levels of radiation, it is necessary to develop a radioprotector *and* mitigator that can protect against these syndromes. Ideally, such a compound should have a convenient mode of delivery and have low toxicity. Unfortunately there are no radioprotectors or mitigators that have been approved for use in humans for preventing or treating the effects of acute radiation exposure. One agent used to reduce the toxicity of radiation therapy is Amifostine (Ethyol[R]), previously known as WR-2721 [42, 43, 44]. Developed by the US Army Anti-Radiation Drug Development Program, Amifostine works thanks to a thiol compound that scavenges free radicals thereby reducing the levels of oxidative radicals. While it has been shown in studies in rats that Amifostine has some radioprotective effect, there are a number of limitations that include:

1. Narrow time window of administration. To have a radioprotective effect, Amifostine must be administered within 15–30 minutes before radiation exposure.
2. It has only been tested intravenously, although other routes may be possible.
3. Side effects. These include vomiting, nausea, and hypotension [42, 43, 44]. Not ideal for a crew of astronauts, although this would be better than suffering the effects of ARS.

Superoxide dismutase (SOD) has been subject to investigation for the transgene's ability to protect tissues against injury following radiation exposure. In mouse models, administration of this transgene did confer some protection against ulceration, and in mice that were fed a diet rich in antioxidants and administered (SOD), lifespans were increased. Genistein is a soy isoflavone that has been used as an anticancer agent [45, 46]. It works by protecting bone marrow progenitor cells and reducing inflammation in tissues [45, 46].

Captopril meanwhile was originally developed to treat hypertension but has since been investigated as a potential radiation countermeasure for the pulmonary and hematopoietic systems [47]. How Captopril works is not completely understood, but research has shown that it blocks radiation-induced hematopoietic syndrome and reduces inflammation.

DBIBB was first highlighted following a study by Gábor Tigyi published in Chemistry and Biology in 2015. In Tigyi's study, DBIBB was shown to increase survival in mice exposed to radiation even after treatment had been administered 3 days after exposure. In previous research, Dr. Tigyi and his colleagues had discovered that a molecule (lysophosphatidic acid or LPA) generated during blood clotting, activates a receptor (called LPA$_2$) that protects against cell death caused by radiation. In this research, scientists had also identified a compound similar to LPA that protected mice against radiation exposure. The problem with this compound was that it did not target the LPA2 receptor, and it was not potent enough to be used as a pharmacological countermeasure. So the researchers refined their study and engineered a more potent version of the LPA$_2$ receptor and dubbed it DBIBB. They then tested this compound in a mouse study and found that DBIBB increased the survival of radiation-exposed cells and protected DNA. In the study, the group of mice that were not treated with DBIBB had a 20% survival rate, whereas the mice treated with DBIBB had a 93 percent survival rate. The next step was to test DBIBB on human hematopoietic progenitor cells. These cells were subject to radiation before being treated with DBIBB. In this study, DBIBB significantly increased the survival of the cells. While the study was one of a kind, the results suggested that DBIBB could be the first radiomitigator.

Dietary Antioxidant Supplementation

Space radiation induces oxidative stress in cells, so it isn't surprising that scientists have suggested astronauts might combat the effects of this stress by taking antioxidants. Oxidative stress occurs when there is a greater amount of prooxidants (radiation is a prooxidant) than antioxidants, and it is hypothesized that the use of antioxidants might counteract this imbalance [48]. Scientists supporting this hypothesis argue that given the level of oxidative stress astronauts will be exposed to during exploration class missions, their intake of antioxidant vitamins will need to be significantly higher than recommended dietary allowances (RDAs).

One study that tested the "antioxidant as a radiation countermeasure" hypothesis, investigated an antioxidant supplement that contained a concoction of several antioxidant agents (ascorbic acid, co-enzyme Q10, α-lipoic acid, L-Selenomethionine, *N*-acetyl cysteine, and vitamin E succinate) that were expected to reduce radiation-induced oxidative stress. The supplement was administered to mice at a weight basis equivalent to humans. One group of mice was then irradiated at the NASA Space Radiation Laboratory (NSRL), while another control group remained radiation free. Following their test, both groups of mice were examined daily for 2 years for signs of toxicity such as ataxia, lack of grooming, weakness, anorexia, convulsions, twitching, tremors, bleeding, discharges, swelling, or labored respiration. Then, at the end of the 2-year period, the experimental and control groups were examined to measure any differences between the two groups. Since there were no statistically significant differences between the different diet groups, scientists were forced to conclude that antioxidant supplementation did not prevent the debilitating effects of radiation exposure. But there was some positive news, because a more detailed analysis of the results revealed that antioxidant supplementation did prevent the more aggressive manifestations of radiation exposure such as malignant lymphomas and rare tumors.

Nicotinamide Mononucleotide

Nicotinamide mononucleotide (NMN) is a much-hyped antiaging drug that has been developed by scientists in Australia and the United States. NMN works by promoting DNA repair and could therefore help protect astronauts from radiation. NMD works by increasing levels of the oxidized form of nicotinamide adenine dinucleotide (NAD+), a chemical present in cells. NAD+ works by regulating protein interactions that help repair DNA, which is why NAD+ supplements have been very popular, although there has been little evidence that supports they have any antiaging effect. NMN on the other hand works so well that the scientists who performed the research are thinking of taking the drug themselves! In these studies, mice that were fed NMN supplements lived 20 percent longer than mice who were not fed the supplement. Of course, human trials have to be conducted, and assuming those trials are successful, the drug will need to be approved by the US Food and Drug Administration.

How does NMN work? As we age, our body's ability to repair itself becomes less and less efficient because the amount of NAD+ present in the cells declines and declines even more in those exposed to radiation. The theory is if you can increase the amount of NAD+ in the cells, you can enhance DNA repair. And the way you increase the amount of NAD+ in the cells is by adding a booster – NMN – that enhances the ability of cells to repair DNA. In some studies that have tested this theory, the NMN not only increased the cells' ability to repair DNA but actually reversed existing genetic damage. And since it is predicted that about five percent of all the cells in an astronaut's body will die during a roundtrip to Mars, NMN has caught the attention of scientists searching for ways to protect crewmembers during these missions.

Granulocyte Colony-Stimulating Factor

As discussed in the previous chapter, one of the syndromes of ARS is hematopoietic syndrome. This syndrome is characterized by a drop in the number of blood cells. This means there is a reduction in the number of neutrophils, which is important as these cells represent the first line of immune defense. As the numbers of neutrophils fall, the risk of infection increases. Neutrophils are produced in the bone marrow from hematopoietic stem cells (HSCs), which give rise to multipotent progenitors (MPPs). The MPPs divide into mature blood cells via processes involving several regulators that are needed for maintaining homeostasis in the cells. One of the key factors in the division, or differentiation, is granulocyte colony-stimulating factor (G-CSF).

Under normal conditions, most mature neutrophils stay in the bone marrow, and only two percent are released into the bloodstream as mature neutrophils. Once in the bloodstream, the differentiated neutrophils search for signs of infection, and if infection is detected, neutrophil chemoattractants are secreted, which triggers the production of G-CSF (during infection, circulating neutrophils may increase by ten times the normal level).

As discussed earlier, when a person is exposed to high levels of ionizing radiation, neutrophil levels fall, resulting in neutropenia. But hypothetically, if it was possible to stimulate neutrophil levels by adding G-CSF, then perhaps neutropenia could be avoided and infection rates reduced. Such studies have been performed using pegfilgrastim (Neulasta®) which is a recombinant form of G-CSF and filgrastim (Neupogen®). In one study, mice were irradiated to 2 Gy, and their neutrophil counts monitored for 30 days after exposure. Not surprisingly, their neutrophil counts decreased significantly compared with the control group which had not been irradiated. But when a second group of irradiated mice was administered filgrastim, the neutrophil counts returned to normal levels within 2 days. In another part of the study, pegfilgrastim was administered to a control group of unirradiated mice, a procedure that boosted neutrophil counts by 15 times. Whether these effects would be observed in astronauts exposed to organ doses of 2 Gy is unknown, but if filgrastim and pegfilgrastim can show similar effects in humans, the compounds may represent a mild countermeasure to radiation exposure.

Despite myriad research studies, there are still no radioprotectors or mitigators available for long-duration crews. There are some weak mitigators such as vitamin E derivatives, and there are compounds being tested on rodents that may be considered as weak radioprotectors, but none of these have been tested in human studies. So what are the ideal characteristics of the ideal radioprotector/mitigator?

1. A weak radioprotector or mitigator is of little use for Mars-bound crews because of the ever-present danger of solar particle events and the constant exposure to GCR. A crew exposed to a high dose of radiation will need a radioprotector/mitigator capable of blocking radiation-induced mutagenesis and carcinogenesis.
2. These agents will need to be effective for the first 24 hours (or longer) following exposure.
3. They should have a convenient mode of administration – intramuscular or subcutaneous.

Psychological Countermeasures

A happy astronaut is a productive astronaut, but maintaining well-being as missions become longer and longer requires enhanced psychological support. This requirement was recognized in the Shuttle era when NASA began providing these services in an effort to minimize stress and promote well-being and performance. This support includes everything from monitoring cognitive function and psychological

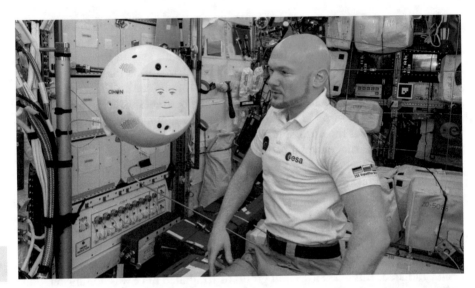

■ **Fig. 7.17**   The $6,000,000 – that's right, six million dollars! – Crew Interactive Mobile companion, or CIMON, is a basketball-sized AI robot. It's not exactly HAL. Not by a long shot; in 2018, the original CIMON (it has since been upgraded to CIMON-2) gained notoriety when it refused to play ESA astronaut Alexander Gerst's favorite song. If that wasn't enough, the obstreperous 5 kilogram "bot" jumped rank and accused the astronaut of being mean! (If you don't believe me you can check out the interaction at this link: ▶ https://www.youtube.com/watch? v=XQQbkDqU1V0). This was a bit much really, considering CIMON's purpose was to boost morale! Credit: NASA

health to ensuring crewmembers have time set aside for family and friends. For example, ISS astronauts can avail themselves of regular private psychological conferences (PPCs) every fortnight. The PPC provides an opportunity for the crewmember to raise any concerns over issues such as sleep, fatigue, mood, and family relationships.

NASA astronauts also have their cognitive function assessed via a monthly WinSCAT (Spaceflight Cognitive Assessment Tool for Windows). The WinSCAT, which is also administered post-mission, can best be described as a neurobehavioral battery designed to determine changes in cognition that may be caused by exposure to toxins such as volatile organic compounds (VOCs). In addition to the PPC and the WinSCAT, astronauts are pampered by having access to private video conferences, a family webpage, personalized care packages ferried up on routine cargo flights, and even access to CIMON (■ Fig. 7.17). Another aspect of psychological support is helping astronauts deal with fatigue caused by the spaceflight-induced misalignment of circadian rhythms. To help manage this fatigue, astronauts may be administered hypnotics and alertness medication.

Mars500

To date, psychological support measures practiced in LEO have been mostly successful (CIMON excepted!), but when astronauts venture beyond Earth orbit, many current countermeasures will either be unavailable or severely limited in scope. What will behavioral scientists do? Maybe they will conduct more analogs such as the Mars 500 boondoggle? The State Scientific Center of the Russian Federation conducted the Mars 500 project in Moscow, which comprised three isolation/confinement studies with six crewmembers each: a 14-day pilot study (completed in November 2007), a 105-day pilot study (completed in July 2009), and a 520-day study simulating a mission to Mars (completed in November 2011). The multinational crew of six were similar in age (32 years) and education (e.g., engineers, physicians, military backgrounds) to astronauts/cosmonauts living on the ISS. The confinement (3 June 2010 to 4 November 2011) was conducted in a 550 m^3 pressurized facility with volume and configuration comparable to a spacecraft. Facility modules were equipped with life support systems and an artificial atmospheric environment at normal barometric pressure, and activities simulated the work routine on board the ISS. Work included routine and simulated emergency events and changes in communication modes and time delays between mission days 54 and 470 that would occur in transit to and from Mars. In many ways, Mars 520 featured many ICE features.

The crew lived on a 5-day work cycle, with 2 days off, except for emergency simulations. Dozens of experiments were performed in the disciplines of physiology, biochemistry, immunology, and biology, microbiology, operations and technology, and of course psychology. Social desirability bias was measured, as was sleep quality together with assessment of mood states to determine depression, tension, anger, and confusion among other things. The Mars500 crew completed depression inventories to see how suicidal or irritable they were and also conflict questionnaires to determine when crewmembers argued the most. Once the mission was over, scientists had reams of data to pore over and publish in peer-reviewed journals.

But did they actually learn anything? Well, the scientists discovered that crewmembers exhibited depressive symptoms and some psychological distress, but nothing that hasn't been observed hundreds of times before in polar explorers. There were also several examples of inter-crew differences in coping with the prolonged isolation and confinement of the 17-month mission, but again, this is nothing new. Sleep-wake data revealed insomnia in some crewmembers and consequent escalating errors in psychomotor vigilance performance. Researchers observed this could be detrimental during critical periods of the mission such as docking maneuvers, extravehicular activities, or responding to emergencies. Perhaps, but polar explorers were subject to extended periods of insomnia compounded by the most horrendous conditions imaginable and were still able to deal with critical tasks.

The researchers attempted to justify their research by stating the importance of identifying behavioral and psychological markers that predispose long-duration crewmembers to behavioral and psychosocial reactions to the confinement required for exploration missions. They went on to say that such predictors and biomarkers are needed to select and train crews and that Mars missions will require the "right stuff" for prolonged confinement and isolation. Well, they're right about that, but a trip to the local library might have told the researchers everything they needed to know about a human's capacity to survive in isolation and confinement, without the need to lock a crew up in a tin can for 17 months. Ultimately, the Mars500 analog was extremely limited, not only by its absence of zero gravity, but because of the very generous comfort blanket available to the crew, which could have left the module at any time. En route to Mars, there will be no such comfort blanket. So, based on the results of Mars500, were researchers able to answer the question "Are humans able to endure the confinement of a trip to Mars?" No. But, based on the experiences of Shackleton, Nansen, and Co., humans most certainly are.

Protecting the Immune System

The immune system interfaces with several systems in the body such as the nervous system and the skeletal system, and it is also influenced by factors such as stress, nutrition, and exercise. Not surprisingly, the space environment exerts a significant influence on this system due to all the stressors present such as radiation, confinement, isolation, microgravity, fluid shifts, and demanding work schedules. All these stressors conspire to cause immune dysregulation, as evidenced by altered distribution of leukocytes and altered cytokine profiles in several astronauts [49, 50, 51]. The use of terrestrial analogs such as Antarctic winter-over and closed chamber habitats has helped determine the mechanism of spaceflight effects, but no analog can replicate all inflight variables. It stands to reason that any immune system changes observed during orbital spaceflight can be expected to be observed in deep space missions, and, with missions to the Moon and Mars being planned, the topic of immune system countermeasures assumes ever greater importance.

Health Countermeasures

One general countermeasure that has been around for a while is the Flight Crew Health Stabilization Program (HSP), the purpose of which is to:

> » ...mitigate the risk of occurrence of infectious disease among astronaut flight crews in the immediate preflight period. Infectious diseases are contracted through direct person-to-person contact, and through contact with infectious material in the environment. The HSP establishes several controls to minimize crew exposure to infectious agents. The HSP provides a quarantine environment for the crew that minimizes contact with potentially infectious material. The HSP also limits the number of individuals who come in close contact with the crew. The infection-carrying potential of these primary contacts (PCs) is minimized by educating them in ways to avoid infections and avoiding contact with the crew if they are or may be sick. The transmission of some infectious diseases can be greatly curtailed by vaccinations.

So, quarantine is one step toward protecting the immune system, but quarantine can only do so much. Another step is designing spacecraft with specific controls designed to reduce infections. These controls include HEPA air filters, water filters, contamination resistance surfaces, biocides, and water pasteurization. In addition to this, microbiological monitoring of cargo, air, payloads, and food is performed prior to launch. Another countermeasure that is implemented is nutritional balance. We know from earlier and current missions that astronauts don't eat sufficient calories, resulting in hypocaloric nutrition. This is not good because inadequate nutrient intake may compromise immune function by causing oxidative stress which in turn causes an inflammatory response [52, 53, 54]. And, in deep space missions where astronauts will be exposed to much more radiation, oxidative stress will be increased and alter genetic repair mechanisms which may ultimately cause immune system failure. What can astronauts do? They can increase their intake of fruits and vegetables, because fruit and vegetables are high in carotenoids, flavonoids, and vitamin C, which improve immune function, which in turn results in an increased antioxidant capacity and a reduction in DNA strand breaks.

Supplements

Another approach is to provide astronauts with supplements. For example, vitamin E is a strong antioxidant, while vitamin A provides an immune-boosting effect. Vitamin C meanwhile is very effective in fighting off oxidative damage and also for stimulating cellular functions of the immune system. Then there is vitamin D, which is important for regulating calcium homeostasis. Another potential supplement is polyphenols (e.g., quercetin, catechins), which have antioxidant and anti-inflammatory benefits. In addition to extra vitamins, there is a case to be made for giving astronauts extra omega-3 fatty acids. Omega-3 fatty acids are long-chain, polyunsaturated fatty acids that have been shown to protect from oxidative damage such as that incurred from radiation exposure.

Pharmacological Countermeasures

In addition to supplementation, there are various medications that astronauts may find helpful. For example, beta-blockers may be useful because they have shown to help reduce the risk of fractures. They can also help modulate memory. On Earth at least. In space, it is a different story because there is insufficient knowledge on drug stability and pharmacodynamics. So, while drug therapy may seem an obvious solution to the many problems of space travel, because of the strong interactions of the immune system with other organ systems, it is very difficult to safely administer this category of countermeasures. And it is not just the problem of pharmacodynamics that must be considered. Many terrestrial drugs have many side effects that would compromise astronaut performance. Fosamax®, Fosavance®, Adrovance®, and Aclasta® are all drugs that are used to preserve bone, but each has side effects. Take Fosamax®, for example. Here are just some of the side effects of taking this drug:

» Fosamax can cause serious side effects. Fosamax can cause irritation, inflammation, or ulcers of the esophagus which may sometimes bleed.

Since Fosamax can cause low calcium levels, if you have low blood calcium before you start taking Fosamax, it may get worse during treatment and must be treated before you take Fosamax.

Symptoms of low blood calcium include, spasms; twitches or cramps in your muscles; and numbness or tingling in your fingers, toes, or around your mouth.

Bone, joint, or muscle pain: Fosamax can cause severe bone, joint, or muscle pain.

Fosamax can cause jawbone tissue to break down, exposing the bone and possibly leading to infections, gum lesions and loosened teeth.

Unusual thigh bone fractures: Fosamax can cause fractures in thigh bones. Symptoms of a fracture may include new or unusual pain in your hip, groin, or thigh.

Allergic reactions and asthma: Fosamax can also cause allergic reactions, such as hives or swelling of your face, lips, tongue, or throat.

Exercise

In addition to building bone and maintaining muscle tone, exercise exerts a powerful positive effect on the immune system. For example, regular exercise reduces low-grade inflammation, improves immune responses to influenza, reduces symptoms of upper respiratory tract infections, and improves mucosal immunity. Having said that, excessive exercise can actually impair immune system function, so any exercise regime must be balanced. Moderation is the key.

Vaccination

Why vaccinate astronauts? Well, as we know now, astronauts' immune systems are dysregulated in spaceflight, and this leads to reactivation of certain viruses such as varicella zoster virus. Furthermore, viral shedding as observed in spaceflight may cause illness. So, to counter this problem, NASA vaccinates all its crews with Zostavax.

Key Terms
- Advanced Resistive Exercise Device (ARED)
- As High As Reasonably Safe (AHARS)
- As Low As Reasonably Achievable (ALARA)
- Bone Mass Density (BMD)
- Central Nervous System (CNS)
- Combined Operational Load-bearing External Resistance Device (COLBERT)
- Computed Tomography (CT)
- International Commission on Radiological Protection (ICRP)
- International Space Station (ISS)
- Interim Resistive Exercise Device (IRED)
- Japanese Experimental Module (JEM)

- Medical Operations Requirement Document (MORD)
- National Academy of Sciences (NAS)
- National Council on Radiation Protection (NCRP)
- Occupational Safety and Health Administration (OSHA)
- Permissible Exposure Limit (PEL)
- Relative Biological Effectiveness (RBE)
- Risk of Exposure Induced Death (REID)
- Russian Orbital Segment (ROS)
- Treadmill with Vibration Isolation and Stabilization System (TVIS)
- United States Orbital Segment (USOS)

❓ Review Questions

1. List four exercise countermeasure devices on board the ISS.
2. Describe the daily prescribed exercise routine of astronauts on board the ISS.
3. What is the CEVIS?
4. What is a DSB?
5. What is meant by PEL?
6. What is meant by ALARA and AHARS?
7. What is meant by weighting factor?
8. Explain how a TLD measures radiation.
9. Explain how a TEPC works.
10. What is PADLES?
11. What is Liulin used for?
12. Explain why magnetic shielding is not a practical solution for protecting astronauts from radiation.
13. Explain why electrical shielding is not a practical solution for protecting astronauts from radiation.
14. What is spallation?
15. How does NMN work to protect astronauts from radiation?

References

1. Moore, A. D., Jr., Downs, M. E., Lee, S. M., Feiveson, A. H., Knudsen, P., & Ploutz- Snyder, L. (2014). Peak exercise oxygen uptake during and following long-duration spaceflight. *Journal of Applied Physiology, 117*, 231–238.
2. Hackney, K. J., Downs, M. E., & Ploutz-Snyder, L. (2016). Blood flow restricted exercise compared to high load resistance exercise during unloading. *Aerospace Medicine and Human Performance, 87*, 688–696.
3. NASA SP-12. Results of the Third U.S. Manned Orbital Spaceflight. 1962.
4. Hayes, J. C., Thornton, W. E., Guilliams, M. E., Lee, S. M. C., MacNeill, K., & Moore, A. D., Jr. (2013). Exercise: developing countermeasure systems for optimizing astronaut performance in space. In W. H. Paloski, D. Risin, & P. Stepaniak (Eds.), *Biomedical Results of the Space Shuttle Program* (pp. 289–313). Washington, DC: US Government Printing Office.
5. Levine, B. D., Lane, L. D., Watenpaugh, D. E., Gaffney, F. A., Buckey, J. C., & Blomqvist, C. G. (1996). Maximal exercise performance after adaptation to microgravity. *Journal of Applied Physiology, 81*, 686–694.

6. Hackney, K. J., Scott, J. M., Hanson, A. M., English, K. L., Downs, M. E., & Ploutz-Snyder, L. L. (2015). The astronaut-athlete: optimizing human performance in space. *Journal of Strength and Conditioning Research, 29*, 3531–3545.

7. Loehr, J. A., Guilliams, M. E., Petersen, N., Hirsch, N., Kawashima, S., & Ohshima, H. (2015). Physical training for long-duration spaceflight. *Aerospace Medicine and Human Performance, 86*, A14–A23.

8. Korth, D. W. (2015). Exercise countermeasure hardware evolution on ISS: the first decade. *Aerospace Medicine and Human Performance, 86*, A7–A13.

9. Cavanagh, P. R., Genc, K. O., Gopalakrishnan, R., Kuklis, M. M., Maender, C. C., & Rice, A. J. (2010). Foot forces during typical days on the international space station. *Journal of Biomechanics, 43*, 2182–2188.

10. https://ntrs.nasa.gov/archive/nasa/casi.ntrs.nasa.gov/20060013245.pdf

11. English, K. L., Lee, S. M. C., Loehr, J. A., Ploutz-Snyder, R. J., & Ploutz-Snyder, L. L. (2015). Isokinetic strength changes following long-duration spaceflight on the ISS. *Aerospace Medicine and Human Performance, 86*, A68–A77.

12. Moore, A. D., Lee, S. M. C., Stenger, M. B., & Platts, S. H. (2010). Cardiovascular exercise in the U.S. space program: past, present and future. *Acta Astronautica, 66*, 974–988.

13. Paoli, A., Gentil, P., Moro, T., Marcolin, G., & Bianco, A. (2017). Resistance training with single vs. multi-joint exercises at equal total load volume: effects on body composition, cardiorespiratory fitness, and muscle strength. *Frontiers in Physiology, 8*, 1105.

14. Weston, M., Taylor, K. L., Batterham, A. M., & Hopkins, W. G. (2014). Effects of low-volume high-intensity interval training (HIT) on fitness in adults: a meta-analysis of controlled and non-controlled trials. *Sports Medicine, 44*, 1005–1017.

15. Baker, J. S., Davies, B., Cooper, S. M., Wong, D. P., Buchan, D. S., & Kilgore, L. (2013). Strength and body composition changes in recreationally strength-trained individuals: comparison of one versus three sets resistance training programmes. *BioMed Research International, 2013*, 615901.

16. Goetchius, L., Scott, J., English, K., Buxton, R., Downs, M., Ryder, J., et al. (2019). High intensity training during spaceflight: results from the SPRINT study. In *Proceedings of the NASA Human Research Program Investigators' Workshop 'Human Exploration and Discovery: The Moon, Mars and Beyond!* (pp. 22–25). Galveston, TX: GICC.

17. Laughlin, M. S., Guilliams, M. E., Nieschwitz, B. A., & Hoellen, D. (2015). Functional fitness testing results following long-duration ISS missions. *Aerospace Medicine and Human Performance, 86*, A87–A91.

18. Matsuo, T., Ohkawara, K., Seino, S., Shimojo, N., Yamada, S., Ohshima, H., et al. (2012). An exercise protocol designed to control energy expenditure for long-term space missions. *Aviation, Space, and Environmental Medicine, 83*, 783–789.

19. Scott, J. P. R., Green, D. A., & Weerts, G. (2018). The influence of body size and exercise countermeasures on resources required for human exploration missions. In *Proceedings of the 39th Annual Meeting of the International Society of Gravitational Physiology (ISGO) & ESA Space Meets Health Initiative* (pp. 18–22). Noordwijk: ESA.

20. Cogoli, A. (1993). The effect of space flight on human cellular immunity. *Environmental Medicine, 37*, 107–116.

21. National Council on Radiation Protection and Measurements. Information needed to make radiation protection recommendations for space missions beyond Low-Earth Orbit. NCRP Report No. 153, Bethesda MD, 2006.

22. Cucinotta, F. A., Kim, M. H., Willingham, V., & George, K. A. (2008). Physical and biological organ dosimetry analysis for International Space Station astronauts. *Radiation Research, 170*, 127–138.

23. ICRP Publication 60, Recommendations of the International Commission on Radiological Protection, Pergamon Press Inc., 1991. ICRP Publication 103 The 2007 Recommendations of the International Commission on Radiological Protection. Annals of the ICRP 37/2-4, 2007. 8. National Academy of Sciences, NAS, National Research Council, Radiation Protection Guides and Constraints for Space-Mission and Vehicle-Design Studies Involving Nuclear System, Washington D.C., 1970.

24. Zeitlin, C., Hassler, D. M., Cucinotta, F. A., Ehresmann, B., Wimmer-Schweingruber, R. F., Brinza, D. E., et al. (2013). Measurements of energetic particle radiation in transit to Mars on the Mars science laboratory. *Science, 340*, 1080–1084.

25. Hassler, D. M., Zeitlin, C., Wimmer-Schweingruber, R. F., Ehresmann, B., Rafkin, S., Eigenbrode, J. L., et al. (2014). Mars' surface radiation environment measured with the Mars science laboratory's curiosity rover. *Science, 343*, 1244797.

26. Kodaira, S., Kawashima, H., Kitamura, H., Kurano, M., Uchihori, Y., Yasuda, N., Ogura, K., Kobayashi, I., Suzuki, A., Koguchi, Y., Akatov, Y. A., Shurshakov, V. A., Tolochek, R. V., Krasheninnikova, T. K., Ukraintsev, A. D., Gureeva, E. A., Kuznetsov, V. N., & Benton, E. R. (2013). Analysis of radiation dose variations measured by passive dosimeters onboard the International Space Station during the solar quiet period (2007–2008). *Radiation Measurements, 49*, 95–102.

27. Labrenz, J., Burmeister, S., Berger, T., Heber, B., & Reitz, G. (2015). Matroshka DOSTEL measurements onboard the International Space Station (ISS). *Journal of Space Weather and Space Climate, 5*, 10.

28. Reitz, G., Berger, T., Bilski, P., Facius, R., Hajek, M., Petrov, V. P., Puchalska, M., Zhou, D., Bossler, J., Akatov, Y. A., Shurshakov, V. A., Olko, P., Ptaszliewicz, M., Bergmann, R., Fugger, M., Vana, N., Beaujean, R., Burmeister, S., Bartlett, D., Hager, L., Palfalvi, J. K., Szabó, J., O'Sullivan, D., Kitamura, H., Uchihori, Y., Yasuda, N., Nagamatsu, A., Tawara, H., Benton, E. R., Gaza, R., McKeever, S. W., Sawakuchi, G., Yukihara, E. G., Cucinotta, F. A., Semones, E., Zapp, E. N., Miller, J., & Dettmann, J. (2009). Astronaut's Organ Doses Inferred from Measurements in a Human Phantom Outside the International Space Station. *Radiation Research, 171*(2), 225–235.

29. Semkova, J., Koleva, R., Shurshakov, V., Benghin, V., St, M., Kanchev, N., et al. (2007). Status and calibration results of Liulin-5 charged particle telescope designed for radiation measurements in a human phantom onboard the ISS. *Advances in space Research, 40*, 1586–1592.

30. Zhou, D., Semones, E., & Weyland, J. (2007). Radiation measured with TEPC and CR-39 PNTDs in low earth orbit. *Advances in Space Research, 40*(11), 1571–1574.

31. Dachev, T. P., Spurny, F., & Ploc, O. (2011). Characterization of radiation environment by Liulin type spectrometers. *Radiation Protection Dosimetry, 144*, 680–683.

32. Dachev, T. P., Semkova, J. V., Tomov, B. T., Matviichuk, Y. N., Maltchev, P. G. S., Koleva, R., et al. (2015). Overview of the Liulin type instruments for space radiation measurement and their scientific results. *Life Sciences in Space Research, 4*, 92–114.

33. Casolino, M., Bidoli, V., Furano, G., Minori, M., Morselli, A., Narici, L., et al. (2002). The Sileye-3/Alteino experiment on board the International Space Station. *Nuclear Physics B, 113*, 71–78.

34. Narici, L., Belli, F., Bidoli, V., Casolino, M., De Pascale, M. P., Di Fino, L., et al. (2004). The ALTEA/Alteino projects: studying functional effects of microgravity and cosmic radiation. *Advances in Space Research, 33*, 1352–1357.

35. Narici, L., Bidoli, V., Casolino, M., De Pascale, M. P., Furano, G., Morselli, A., et al. (2003). ALTEA: anomalous long-term effects in astronauts. A probe on the influence of cosmic radiation and microgravity on the central nervous system during long flights. *Advances in Space Research, 31*, 141–146.

36. Semkova, J., Koleva, R., Maltchev, S., Benghin, V., Shurshakov, V., Chernykh, I., et al. (2008). Preliminary results of Liulin-5 experiment for investigation of the dynamics of radiation doses distribution in a human phantom aboard the International Space Station. *Comptes Rendus de l Academie Bulgare des Sciences, 61*, 787–794.

37. Wilson, J. W., Thibeault, R. C., Cucinotta, F. A., Shinn, M. L., Kim, M. H., Kiefer, R., & Badavi, F. F. (1995). Issues in protection from galactic cosmic rays. *Radiation and Environmental Biophysics, 34*, 217.

38. Zeitlin, C., Guetersloh, S., Heilbronn, L., & Miller, J. (2005). Shielding and Fragmentation Studies. *Radiation Protection Dosimetry, 116*, 123–124.

39. Durante, M., George, K., Gialanella, G., Grossi, G., La Tessa, C., Manti, L., Miller, J., Pugliese, M., Scampoli, P., & Cucinotta, F. A. (2005). cytogenetic effects of high-energy iron ions: dependence on shielding thickness and material. *Radiation Research, 164*, 571–576.

40. Harrison, C., Weaver, S., Bertelsen, C., Burgett, E., Hertel, N., & Grulke, E. (2008). Polyethylene/ Boron Nitride Composites for Space Radiation Shielding. *Journal of Applied Polymer Science, 109*, 2529–2538.
41. Estevez, J. E., Ghazizadeh, M., Ryan, J. G., & Kelkar, A. D. (2014). Simulation of hydrogenated boron nitride nanotubes mechanical properties for radiation shielding applications. *International Journal of Engineering Science, 8*(1), 63–67.
42. Kouvaris, J. R., Kouloulias, V. E., & Vlahos, L. J. (2007). Amifostine: the first selective-target and broad-spectrum radioprotector. *Oncologist, 12*(6), 738–747.
43. van der Vijgh, W. J., & Peters, G. J. (1994). Protection of normal tissues from the cytotoxic effects of chemotherapy and radiation by amifostine (Ethyol): preclinical aspects. *Seminars in Oncology, 21*(5) Suppl 11, 2–7.
44. Bourhis, J., Blanchard, P., Maillard, E., Brizel, D. M., Movsas, B., Buentzel, J., et al. (2011). Effect of amifostine on survival among patients treated with radiotherapy: a meta-analysis of individual patient data. *Journal of Clinical Oncology, 29*(18), 2590–2597.
45. Verdrengh, M., Jonsson, I. M., Holmdahl, R., & Tarkowski, A. (2003). Genistein as an anti-inflammatory agent. *Inflammation Research, 52*(8), 341–368.
46. Landauer, M. R., Srinivasan, V., & Seed, T. M. (2003). Genistein treatment protects mice from ionizing radiation injury. *Journal of Applied Toxicology, 23*(6), 379–385.
47. Davis, T. A., Landauer, M. R., Mog, S. R., Barshishat-Kupper, M., Zins, S. R., Amare, M. F., et al. (2010). Timing of captopril administration determines radiation protection or radiation sensitization in a murine model of total body irradiation. *Experimental Hematology, 38*(4), 270–281.
48. Rizzo, A. M., Corsetto, P. A., Montorfano, G., Milani, S., Zava, S., Tavella, S., et al. (2012). Effects of long-term space flight on erythrocytes and oxidative stress of rodents. *PLoS One, 7*, e32361.
49. Sonnenfeld, G. (1994). Effect of space flight on cytokine production. *Acta Astronautica, 33*, 143–147.
50. Konstantinova, I. V., Rykova, M. P., Lesnyak, A. T., & Antropova, E. A. (1993). Immune changes during long-duration missions. *Journal of Leukocyte Biology, 54*, 189–201.
51. Crucian, B., Stowe, R. P., Mehta, S., Quiriarte, H., Pierson, D., & Sams, C. (2015). Alterations in adaptive immunity persist during long-duration spaceflight. *NPJ Microgravity, 1*, 15013.
52. Mermel, L. A. (2013). Infection prevention and control during prolonged human space travel. *Clinical Infectious Diseases, 56*, 123–130.
53. Crucian, B., Babiak-Vazquez, A., Johnston, S., Pierson, D. L., Ott, C. M., & Sams, C. (2016). Incidence of clinical symptoms during long-duration orbital spaceflight. *International Journal of General Medicine, 9*, 383–391.
54. Guéguinou, N., Huin-Schohn, C., Bascove, M., Bueb, J.-L., Tschirhart, E., Legrand-Frossi, C., et al. (2009). Could spaceflight-associated immune system weakening preclude the expansion of human presence beyond Earth's orbit? *Journal of Leukocyte Biology, 86*, 1027–1038.

Suggested Reading

Chang-Díaz, F., Seedhouse, E. (2017). To Mars and Beyond, Fast! How Plasma Propulsion will Revolutionize Space Exploration. Springer-Praxis.
Cucinotta, F. A., Kim, M. H., Willingham, V., & George, K. A. (2008). Physical and biological organ dosimetry analysis for International Space Station astronauts. *Radiation Research, 170*, 127–138.
Moore, A. D., Lee, S. M. C., Stenger, M. B., & Platts, S. H. (2010). Cardiovascular exercise in the U.S. space program: past, present and future. *Acta Astronautica, 66*, 974–988.
Seedhouse, E. (2018). *Space Radiation and Astronaut Safety*. Springer Brief.

Growing Food in Space

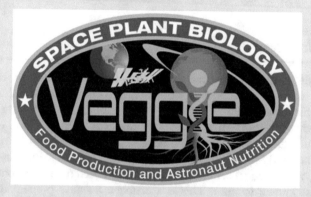

Credit: NASA

Contents

© Springer Nature Switzerland AG 2020
E. Seedhouse, *Life Support Systems for Humans in Space*,
https://doi.org/10.1007/978-3-030-52859-1_8

🔬 Learning Objectives
- Explain the process of germination
- Distinguish between conduits and tubules
- Explain what is meant by phototropism, gravitropism, and dietary fatigue
- Explain how plant pillows work
- Describe the work conducted in the Veggie facility
- Describe the four loops of the MELiSSA project

Introduction

As long as there have been astronauts, there have been processed, prepackaged space rations. Processed fruits, prepackaged nuts, irradiated shrimp cocktail, sterilized chicken stew, fluffernutter (a favorite food item of Sunita Williams; I had no idea what this food – if it indeed can be categorized as a food – was until I showed my students a video wherein Sunita explained the items stored in the pantry on International Space Station), you name it. These space rations are heated, freeze-dried, irradiated, thermostabilized, and subjected to just about every processing process known to man. This is a problem because prepackaged space foods are sometimes deficient in nutrients (e.g., potassium and vitamin K), and whatever nutrients they do contain may degrade over time (e.g., vitamin B1 and vitamin C) [1]. But now, thanks to plant growth experiments being flown on orbit, it may soon be possible for crop production to be integrated into life support systems of future spacecraft, thereby allowing astronauts to supplement a stored and packaged diet with freshly grown vegetables.

Plants on Earth: A Primer

There are more than 300,000 plant species, ranging in size from microscopic to the largest living organisms on the planet. In common with all organisms, plants need energy, nutrients, air, and water. What follows is a short primer on some key plant characteristics. The green color is thanks to a pigment known as *chlorophyll*, which helps plants capture light energy via photosynthesis. During this process, leaves extract carbon dioxide to store energy that is used to help plants grow. As this process is taking place, oxygen is released.

The largest category of plants is angiosperms, which are those plants with roots, stems, leaves, and flowers. The roots serve two very important functions, one of which is to anchor the plant and the other of which is to absorb water and nutrients. Stems on the other hand serve as channels through which nutrients and water are directed between the roots and leaves, which is where gas exchange (the carbon dioxide and oxygen mentioned earlier) occurs, thanks to pores in the leaves. The flowers are the reproductive part of the plant, and it is here that pollination takes place. In some plants, this pollination leads to the formation of fruit – apples, oranges, peas, acorns, and grapes, for example. These fruits serve myriad important roles, one of which is seed dispersal.

It is the recycling and food properties of plants that make them such a godsend for engineers designing closed life support systems. You see, without plants, any manned mission to Mars or beyond is a no-go. Period. Nevertheless, integrating the growing of plants into a life support system is a fiendishly difficult enterprise. Just ask the MELiSSA engineers and scientists at the University of Barcelona. These talented individuals have been working for *30 years*, and still they have not been able to solve the puzzle of growing plants in a closed LSS. In order to do so, scientists must replicate a terrestrial environment in space. To give you an idea of just how difficult that is, it is necessary to understand the intricacies of the plant-growing process. So here goes.

Germination

The seeds of flowering plants comprise a protective coat, an embryo, and a stored food. The embryo is a new plant that stays dormant until suitable conditions occur for germination. At one end of the embryo is a feature called the *radicle*, and it is the radicle that ultimately develops into the root. At the other end is the *hypocotyl*, which eventually forms the stem and the leaves. The food? Well, that can be stored either in the seed leaves or around the embryo. Once conditions are suitable for germination, the seed and the embryo take in water, the plant begins to utilize its fuel, and growth begins. In some cases, germination must be triggered by very specific conditions such as light at very specific wavelengths or a number of days with very little temperature variation. During germination, our new plant grows a single root to collect water and to serve as an anchor. After a while, the radicle becomes the primary root. But not all roots are alike. Some roots, such as those that belong to carrots, form fat roots, whereas other plants grow thin roots. Then, as growth continues, leaves begin to form, and the process of photosynthesis begins, but before we look more closely at this process, it is instructive to know a little more about roots and stems.

Roots and Stems

First of all, roots do a whole lot more than just anchor a plant. They absorb water and absorb nutrients, and there are some plants whose roots house bacteria capable of trapping nitrogen and making it available to the plant. On Earth, most roots grow continuously and follow the path of least resistance. The direction and proliferation of the root system is largely determined by factors such as water, oxygen availability, and nutrients in the soil. Water and nutrients are absorbed through tubes known as "root hairs," whose purpose is to increase the surface area for water and nutrients to make their way into the root system. For the rest of the plant, water and nutrients are transported through the vascular system, which comprises a network of *tubules* (which transport water and nutrients) and *conduits* (which carry the products of photosynthesis). Many plants have roots and stems that are reinforced to allow them to survive long periods of time.

Now, you may be wondering how roots know which direction to grow given that plants do not have a nervous system. Well, plants can still *sense*, thanks to special receptor molecules that detect and respond to changes in light. Once these changes are detected, signaling pathways are triggered and plants react accordingly. These reactions, which are basically a series of chemical signals, are how plants communicate with each other, and it is how roots know which direction to grow. But it isn't just roots that must know what to do, because there must also be triggers for flowering and seed germination.

How are these processes triggered? The answer is light. A plant's response to light is called *phototropism*, which triggers biochemical responses that ultimately determine how fast or how slow a plant grows. Because stems grow toward light, they are said to have a *positive phototropic response*, whereas roots, which only have a very weak response to light, are said to have a *negative phototropic response*. At this point, it is also important to note that different plants respond to different colors of light, so this is yet another variable that must be considered and monitored by astronaut gardeners. Of course the most powerful trigger for root growth and root orientation is gravity. Gravity is a much stronger stimulus for root orientation than light is for stem growth. The stimulus gravity provides for root orientation is known as *gravitropism*. Roots are *positively gravitropic* and stems are *negatively gravitropic*.

Veggie

"Our plants aren't looking good." Those were the words of Scott Kelly as he tweeted the latest plant-growing update from the International Space Station (ISS) in 2015. The subject of the tweet was a rather pitiful-looking bunch of baby zinnias that had curled up and given up the ghost. The killer? Mold, apparently. Gardeners here on Earth encounter similar problems, which usually prompts a trip to the local gardening shop. But that won't be an option on a trip to Mars. No local nurseries in deep space last time anybody checked. As we know by now, the biggest gap in closing the loop in the LSS is food production; interplanetary astronauts will need space gardens if they are to have any chance of surviving a reasonable distance from LEO. One step (and there are many!) toward achieving that goal is figuring out how to prevent mold from destroying crops. Fortunately, one system is already helping scientists figure out the orbital plant-growing business, and its name is Veggie (◘ Fig. 8.1).

Plant Pillows

Veggie is a plant growth test facility that currently resides on board the ISS. It's a simple low-power system that helps astronauts test plants for eventual consumption. Built by Orbitec in Madison, Wisconsin, thanks to a NASA Small Business Innovation Research (SBIR) Program, the Veggie system flew to ISS on SpaceX's CRS-3 mission in April 2014 [2]. Shortly after Veggie's arrival, astronauts Rick Mastracchio and Steve Swanson plugged the system into the Columbus module.

Fig. 8.1   Veggie. Credit: NASA

Fig. 8.2   Plant pillows. Credit: NASA

How does Veggie work? Passive wicking is used to provide water to the plants as they grow in specially designed plant pillows (■ Fig. 8.2) that contain the growth media, which is basically a special type of clay.

Fertilizer is released into the plant pillows, as is water, which is injected through a valve. At the beginning of each plant test, each plant pillow receives up to three seeds (two are backups in case of germination failure). These seeds are oriented so the plant grows upward and the roots grow downward. The plants then begin to grow under the daily supervision of astronauts. Simultaneously, an identical crop is grown on Earth. Some crops grow well, and some not so well. This isn't surprising because there are so many variables that influence plant growth: oxygen, carbon dioxide, water, temperature variations, humidity variations, and the list goes on [3, 4]. And then there is the absence of gravity, which causes problems because of the lack of convective flow.

Over the years, astronauts have grown more than half a dozen types of leafy greens in Veggie, and more than 100 crops have been tested on the ground. The first crop, VEG-01, was lettuce planted by Steve Swanson in May 2014. The crop comprised one set of six plant pillows, planted with red romaine lettuce seeds. After 33 days, one plant pillow was reported as having not germinated, and two other plant pillows were lost due to water stress. Ultimately, three healthy plants were the result. But these weren't eaten by the crew. Instead, the plants were returned to Earth for testing and analysis.

The second Veggie experiment, VEG-01 B, began in July 2015. This time it was Scott Kelly's turn to work as a gardener together with some help from Kjell Lindgren. Once again, six plant pillows were planted, each containing red romaine lettuce seeds. After 33 days, the crew were given permission to harvest half the crops and taste the lettuce, marking the first time crops had been grown using NASA hardware and then consumed. Now you may think eating some lettuce in space is no big deal, but think about it again after eating bland rehydrated fare for several months. Not that there is anything wrong with the variety of food in the ISS pantry, but astronauts do succumb to *dietary fatigue*, which is why lettuce was a welcome change. The other half of the crops were returned to Earth for microbial analysis.

The third crop, VEG- 01 C, comprised a set of plant pillows planted with zinnia seed. Zinnia was chosen because it has a longer growth period than lettuce. This third crop-growing exercise was not without its problems, one of which was water leaking from the wicks that held the seeds. This resulted in moisture seeping from the leaves, which then began to curl up as mold set in. In an attempt to save the zinnias, the astronauts turned the airflow up to the high setting, but it wasn't enough. The leaves began to die. This prompted the astronauts to break out the clippers and begin to snip away the moldy parts. It was a good attempt, but the high fan setting resulted in the plants becoming dehydrated. Not a great situation for a Mars-bound crew, but fortunately this was the ISS. As always, NASA was able to put a positive spin on the whole affair, stating that at least they knew zinnia could survive flood and drought! Ultimately, some zinnias were harvested successfully, and seeds from the space-borne zinnias were later germinated on Earth.

Advanced Plant Growth Habitat

Another plant-growing facility on ISS is the Advanced Plant Growth Habitat (◘ Fig. 8.3), which comprises a 45-cm square self-sufficient laboratory fitted with 180 sensors and an automated watering capability. A temperature regulation system permits air temperatures to be set within 0.5 °C, and the aforementioned sensors relay data about light, humidity, and oxygen levels to the ground. What is the difference between Veggie and the APH you may be thinking? Well, Veggie is being used to determine how and why astronauts can grow a food supply, whereas the APH is being used to quantify the actual circumstances that allow Veggie to grow plants. Since being assembled on orbit in October 2017, the APH, which has more

Fig. 8.3 The Advanced Plant Growth Habitat (APH) with door removed. The plant featured is dwarf wheat. Credit: NASA

than a passing resemblance to a microwave, has been used to test various crops, including *Arabidopsis thaliana* and dwarf wheat.

Research to Date

As was mentioned earlier, whenever a crop is grown in space, an identical one is grown on the ground so nutritional comparisons can be made. So what have the scientists found?

Well, let us take a look at the nutritional quality of that red romaine lettuce (*Lactuca sativa*) to give the plant its proper moniker [5]. Following return to Kennedy Space Center, samples from the experiment were maintained in a −80 °C freezer until analysis. Plant samples were then thawed and the leafy biomass divided for microbiological and chemical analysis. Following microbial analysis, it was found that bacterial counts of the plants not grown in space were lower than the plants that had been grown in space. A screen for foodborne pathogens (such as *E. coli* and *Salmonella*) revealed negative results, but a bacterial screen found the genus *Staphylococcus*, certain strains of which can be pathogenic to humans. When the flight and ground samples were screened for elemental (phosphate, magnesium, zinc) analysis, no significant differences were found, although sodium was found to be higher in the flight lettuce. Scientists also measured the antioxidant capacity of the flight and ground samples but did not observe a significant variance. Some of the differences between the ground and flight samples were attributed to a mix of terrestrial airborne bacteria, poor air circulation (due to a fan malfunction in the Veggie compartment), and different fluid behavior in microgravity. Ultimately, however, the scientists reported that from a microbiological and bacteriological perspective, red romaine lettuce grown in space was safe for human consumption [6].

MELiSSA

» We are creating an artificial ecosystem which uses micro-organisms to process the waste so that we can grow plants. At the bottom is sludge (raw waste) which undergoes anaerobic (without oxygen) fermentation in darkness. Higher up there's light but no oxygen. Higher still there's oxygen and it's possible to transform ammonia to nitrate. At the surface, there's carbon dioxide, oxygen and light. This is where higher plants can thrive.

     Christophe Lasseur, scientist of the MELiSSA Project team

Being able to grow food the crew can eat is one vital step toward a closed LSS, but how do we integrate Veggie (or a similar system) into the LSS? Well, to get an idea of how that might be achieved a discussion of MELiSSA (◘ Fig. 8.4) is instructive. MELiSSA, aka the Micro-Ecological Life Support System Alternative, is a project of the European Space Agency (ESA) and is located at the University of Barcelona in Catalonia, Spain. The endeavor began in 1988 and has been active ever since [7, 8]. The general model is based on the waste loop cycle of a natural lake ecosystem (◘ Fig. 8.5).

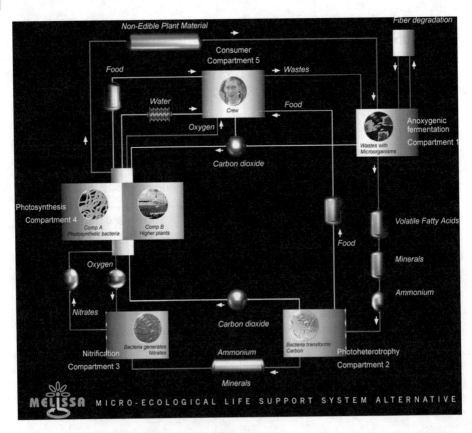

◘ **Fig. 8.4**   The MELiSSA Project. Credit: ESA

Fig. 8.5 MELiSSA is based on an aquatic ecosystem depicted above. Credit: ESA

MELiSSA's Partners and Supporting Subcontractors

The MELiSSA Project is based on a collaborative development program managed by the ESA's ESTEC Thermal and Environmental Control Section (TEC-MCT). The partner institutions are:

- ESA (European Space Agency)
- Université Blaise Pascal, Clermont-Ferrand, France
- Ghent University, Ghent, Belgium
- IPStar B.V., Vught, Netherlands
- SCK-CEN, Belgian Nuclear Research Center
- University of Guelph, Guelph, Ontario, Canada
- Universitat Autònoma de Barcelona, Barcelona, Spain
- Vlaamse Instelling voor Technologisch Onderzoek N.V. (VITO), Mol, Belgium
- Sherpa Engineering S.A., France
- University of Naples Federico II, Naples, Italy
- University of Lausanne, Lausanne, Switzerland
- EnginSoft S.p.A., Mattarello, Italy
- MELiSSA Foundation PS (Private Stichting), Brussels, Belgium

8

■ **Fig. 8.6** This is what a bioreactor looks like. Credit: ESA

In such a system, each output is converted into a new input, thanks to the natural processes occurring in the lake. These processes take place, thanks to the interaction of organisms such as bacteria, fungi, and algae. These processes are replicated in the MELiSSA system, thanks to four bioreactors and a higher plant compartment (HPC) [9, 10]. Each bioreactor (■ Fig. 8.6) functions as a recycling system for wastes and by-products that are funneled into the HPC, which is where plants are grown.

MELiSSA Development

The project follows a five-phase approach: basic research and development, preliminary flight experiments, ground and space demonstration, technology transfer, and communication and education. We'll discuss each of these phases in this section.

Basic Research and Development

One of the first goals of the endeavor was the identification of bacterial strains, because each compartment requires specific bacteria for the system to work effectively. It was a difficult task, as deploying the wrong bacteria could result in imbalances in the system. Another challenge was genetic evolution. Bear in mind that the planned LSS will be used for years. In that time, bacteria can evolve in a way that could change the stability of the system. So, in an effort to troubleshoot these issues, the MELiSSA Genetics (MELGEN) project was created, its purpose being to detect and resolve bacterial evolution [11].

A second task of the basic R&D phase was determining the processing capability of the waste compartment, which led to the creation of a waste compartment prototype that was tested using human fecal matter. Then the focus was directed at the HPC. Even today, the challenges of the HPC remain formidable, because even by applying the most advanced and complex mathematical modeling in the world, simulating all the myriad permutations of plant growth variables in an effort to achieve optimal growth output is a difficult endeavor. And when you throw the microgravity variable into the mix, the problem is amplified.

Preliminary Flight Experiments

Once scientists begin to resolve the problems of growing plants on the ground inside a terrestrial closed LSS, those experiments must be replicated in space. That is because, as discussed before, physicochemical and biological processes are affected in microgravity. Another reason these experiments must be flown in space is because of the radiation environment. Terrestrial plants are happy living in an environment that is exposed to just 2 mSv per year, but what happens when these plants and bacteria are exposed to 80 times that dose? [12] To answer these questions, the MELiSSA program flew the rotating wall vessel and random positioning machine on board the ISS.

Ground and Space Demonstration

Following promising results from those experiments, the BIORAT (a misnomer, because the experiment will fly mice, not rats!) experiment will be flown on ISS.

BIORAT – Putting Mice in the Loop

The BIORAT is a testbed for the engineering principles that will be required for a full-scale closed LSS. Obviously it makes sense to test such a system on a small scale first, so the BIORAT will comprise a simple ecosystem that will house small test animals – mice in this case. The focus of BIORAT will be on two of the loops in the MELiSSA system: the photosynthetic reactor and the consumer compartment. In this test, carbon dioxide produced by the mice will be consumed by algae (*Spirulina*) in the photosynthetic process, thereby resulting in the production of oxygen. During the test, variables such as lighting, gas transfer, and oxygen production will be controlled and growth of *Spirulina* will be modified using different light intensities.

In addition to BIORAT, the MELiSSA development path is testing its technology using the MELiSSA Pilot Plant (MPP). The MPP (◘ Fig. 8.7) comprises five compartments: the first of which is for waste degradation, the second for nitrification, the third for air revitalization (which is there the Spirulina go to work), the fourth for food production, and the fifth is where the mock crew (in this case, mice) reside [8].

In their effort to develop the MPP and the other life support technologies, MELiSSA engages in scientific collaborations with other groups that are pushing life support technology to the edge. For example, MELiSSA works with the Concordia, an Antarctic research station that recycles its wastewater. And as the MPP is being ground tested, data is being applied to terrestrial systems such as self-sustainable habitats.

☐ **Fig. 8.7**   The MELiSSA Pilot Plant [8]. Credit: ESA

8

Technology Transfer

MELiSSA's technology transfer partner is SEMiLLA, which was established in 2005. The company is responsible for generating budgets to enable further research. It does this through four spin-off companies: SEMiLLA IPStar, SEMiLLA Health BV, SEMiLLA sanitation BV, and ezCOL BV. One example of a spinoff technology is the use of bacteria to treat water in shipping ballast.

Communication and Education

MELiSSA pursues several education and informational outreach programs and routinely facilitates presentations at international conferences [13]. I was lucky enough to visit the site in the summer of 2018 with some students from Embry-Riddle Aeronautical University, and MELiSSA scientists were kind enough to spend 2 hours presenting the facility to the students.

So what will a space garden look like? One that is integrated into our fictional closed LSS. It will probably look more like Veggie than the APH, and it will be smaller than MELiSSA. A lot smaller!

But there is long way to go before a Mars-bound Veggie 2.0 is flown. Optimal guidelines for plant growth must be developed, and those guidelines must ensure that space-bound gardening systems function reliably, that plants grow substantially and regularly. A thorough assessment of all myriad variables that influence plant growth in space must be conducted. For example, the effects of contamination [14], of bacteria [15, 16], of carbon dioxide [17], and on the crew [18] and the environment [19] must be evaluated. Once these variables have been assessed, the plant-growing systems must be integrated into the LSS, and monitoring systems [20] must be developed and tested. Then and only then can astronauts crunch their way across the cosmos.

Key Terms
- Advanced Plant Growth Habitat (APH)
- European Space Agency (ESA)
- Higher Plant Compartment (HPC)
- International Space Station (ISS)
- Micro-Ecological Life Support System Alternative (MELiSSA)
- MELiSSA Pilot Plant (MPP)
- Small Business Innovation Research (SBIR)

? Review Questions
1. What is the function of chlorophyll?
2. Explain the function of a radicle and a hypocotyl.
3. What is meant by the term *positive phototropic res*ponse?
4. Describe the process of germination.
5. Explain how plant pillows work.
6. What is meant by *dietary fatigue*?
7. Describe the functions of each of the four levels of MELiSSA.
8. What does a bioreactor do?

References

1. Borchers, A. T., Keen, C. L., & Gershwin, M. E. (2002). Microgravity and immune responsiveness: Implications for space travel. *Nutrition, 18*, 890–898.
2. Massa, G. D., Dufour, N. F., Carver, J. A., Hummerick, M. E., Wheeler, R. M., Morrow, R. C., et al. (2017). Veggie hardware validation testing on the International Space Station. *Open Agriculture, 2*, 33–41.
3. Massa, G. D., Newsham, G., Hummerick, M. E., Morrow, R. C., & Wheeler, R. M. (2017). Preparation for the veggie plant growth system on the International Space Station. *Gravitational and Space Research, 5*, 24–34.
4. Massa, G. D., Wheeler, R. M., Morrow, R. C., & Levine, H. G. (2016). Growth chambers on the International Space Station for large plants. *Acta Horticulture, 1134*, 215–222.
5. Oliveira, M., Usall, J., Viñas, I., Anguera, M., Gatius, F., & Abadias, M. (2010). Microbiological quality of fresh lettuce from organic and conventional production. *Food Microbiology, 27*, 679–684.
6. Khodadad, C. L. M., Hummerick, M. E., Spencer, L. E., Dixit, A. R., Richards, J. T., Romeyn, M. W., Smith, T. M., Wheeler, R. M., & Massa, G. D. (2020). Microbiological and nutritional analysis of lettuce crops grown on the International Space Station. *Frontiers in Plant Science, 11*, 199.
7. Perez, J., Montesinos, J. L., & Godia, F. (1999). *Operation of the nitrifying pilot reactor*. Technical Note 37.420. MELISSA. ESTEC/CONTRACT. 11549/95/NL/FG.
8. Albiol, J., Perez, J., Cabello, F., Mengual, X., Montras, A., Masot, S., Camargo, J., & Gòdia, F. (2003). In M. Lobo & C. Lasseur (Eds.), *Leaving and Living with MELiSSA, MELiSSA Pilot Plant, MELiSSA Final Report 2002 Activity, ESA/EWP-2216* (pp. 206–225).

9. Burtscher, C., Fall, P. A., Christ, O., Wilderer, P. A., & Wuertz, S. (1998). Detection and survival of pathogens during two-stage thermophilic/mesophilic treatment of suspended organic waste. *Water Science and Technology, 38*, 123–126.

10. Chachkhiani, M., Dabert, P., Abzianidze, T., Partskhaladze, G., Tsiklauri, L., Dudauri, T., & Godon, J. J. (2004). 16S rDNA characterization of bacterial and archaeal communities during start-up of anaerobic thermophilic digestion of cattle manure. *Bioresource Technology, 93*, 227–232.

11. Koops, H.-P., & Pommerening-Röser, A. (2001). Distribution and ecophysiology of the nitrifying bacteria emphasizing cultured species. *FEMS Microbiology Ecology, 37*, 1–9.

12. Hendrickx, L., De Wever, H., Hermans, V., Mastroleo, F., Morina, N., Wilmotte, A., Janssen, P., & Mergeay, M. (2006). Microbial ecology of the closed artificial ecosystem MELiSSA (Micro-Ecological Life Support System Alternative): Reinventing and compartmentalizing the earth's food and oxygen regeneration system for long-haul space exploration missions. *Research in Microbiology, 157*, 77–86.

13. Lasseur, C., Brunet, J., de Weever, H., Dixon, M., Dussap, G., Godia, F., Leys, N., Mergeay, M., & Van Der Straeten, D. (2010, August). MELiSSA: The European project of closed life support system. European Space Agency, TEC-MMG Keplerlaan 1, 2200 AG Noordwijk, The Netherlands. *Gravitational and Space Biology, 23*(2), 2–8.

14. Heaton, J. C., & Jones, K. (2008). Microbial contamination of fruit and vegetables and the behaviour of enteropathogens in the phyllosphere: A review. *Journal of Applied Microbiology, 104*, 613–626.

15. Dees, M. W., Lysøe, E., Nordskog, B., & Brurberg, M. B. (2015). Bacterial communities associated with surfaces of leafy greens: Shift in composition and decrease in richness over time. *Applied and Environmental Microbiology, 81*, 1530–1539.

16. Leff, J. W., & Fierer, N. (2013). Bacterial communities associated with the surfaces of fresh fruits and vegetables. *PLoS One, 8*, e59310.

17. McKeehen, J. D., Smart, D. J., Mackowiack, C. L., Wheeler, R. M., & Nielsen, S. S. (1996). Effect of CO2 levels on nutrient content of lettuce and radish. *Advances in Space Research, 18*, 85–92.

18. Perchonok, M., Douglas, G., & Cooper, M. (2012). *Risk of performance decrement and crew illness due to an inadequate food system. HRP Evidence Book. NASA Human Research Program.* Available online at: http://humanresearchroadmap.nasa.gov/Evidence/reports/Food.pdf

19. Sublett, W., Barickman, T., & Sams, C. (2018). The Effect of environment and nutrients on hydroponic lettuce yield, quality, and phytonutrients. *Horticulturae, 4*, 48.

20. Yamaguchi, N., Roberts, M., Castro, S., Oubre, C., Makimura, K., Leys, N., et al. (2014). Microbial monitoring of crewed habitats in space-current status and future perspectives. *Microbes Environments, 29*, 250–260.

Suggested Reading

Lasseur, C. (2008, January). *MELiSSA: The European project of a closed life support system.* Research Gate.

8

Future Life Support Concepts

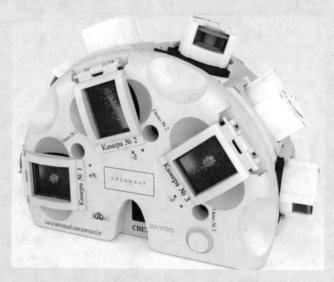

Credit: Private Institution Laboratory for Biotechnological Research, Moscow, Russia, and NASA

Contents

© Springer Nature Switzerland AG 2020
E. Seedhouse, *Life Support Systems for Humans in Space*,
https://doi.org/10.1007/978-3-030-52859-1_9

Learning Objectives

After completing this chapter, you should be able to:

- Explain what is meant by the term *Coriolis force*
- Explain what is meant by the term *torpor*
- List five life support system advantages of having a hibernating crew
- Describe the process of hibernation in animals
- Describe the process of preprocessing, processing, and post-processing in the context of bioprinting
- Describe the structure of a vasculoid
- Distinguish between the function of vasculocytes and respirocytes
- Explain what is meant by defluidization

Introduction

So which life support technologies can we look forward to in the near future? Well, those science fiction aficionados among you will have an idea. Think of all those movies that feature hibernation. As we'll see in this chapter, hibernation isn't science fiction and hasn't been so for years. Therapeutic hypothermia is a standard medical procedure that, with a few tweaks, can be adopted by interplanetary astronauts. Then there is nanotechnology. Think about the problems of muscle atrophy and bone demineralization. Well, nanotechnology may be the solution. And finally, there is artificial gravity. This is a popular science fiction tool that will probably be the toughest life support nut to crack, so let us begin with this.

Artificial Gravity

For years, science fiction aficionados have been inundated with scenes of artificial gravity (*2001: A Space Odyssey*, *Passengers*, *The Martian*, *Elysium*, the list goes on). In addition to images of elegant spaceships providing crews with terrestrial gravity, hopelessly misinformed journalists (when I say misinformed it is because these journalists write their articles under the assumption that artificial gravity exists) have for decades touted this technology as the one that enables us to head to Mars. Headlines pronounce "How that spinning spacecraft from The Martian would work," "Artificial Gravity breaks free from Science Fiction," and "Real Artificial Gravity for SpaceX Starship."

But in 2020, more than 65 years after the great Wernher von Braun described his design for creating artificial gravity, we are still a long, *long* way from realizing this technology. We all know that spending a long time in space is bad for you, and we all know that with 2020 technology, those first steps on Mars will be less of a giant leap and more of an ungainly stagger – and that's assuming those astronaut's legs don't fracture! You see, the problem with baseline humans is that they just haven't evolved for life in space. So if astronauts are ever to venture to Mars and beyond, we need some extreme solutions. Cloning, genetic manipulation, and replicants (□ Fig. 9.1) perhaps? Maybe, but first let's take a look at the possibility of resurrecting the artificial gravity idea.

◘ Fig. 9.1 Author's copy of In Philip K Dick's *Do Androids Dream of Electric Sheep*, replicants are bioengineered beings that have superior strength, speed, agility, and resilience, which make them perfect for doing the jobs that baseline humans can't – such as exploring other planets! Credit: Author

9

So, if we were to engineer some artificial gravity spacecraft, what would it look like? A torus perhaps? In a torus design, thrusters would rotate the structure around its axis, thereby generating centripetal force. This means that any astronaut in the outer part of the structure would experience gravity as the floor of the hull pushed against them. The amount of artificial gravity would depend on two factors: the size of the wheel and the speed of the rotation. The bigger the wheel and the faster it turned, the more pronounced the gravity. This effect is demonstrated in *2001: A Space Odyssey* [1].

A smaller scale version of the von Braun-esque station is the Nautilus-X (◘ Fig. 9.2). Proposed in January 2011 by a brilliant team (Mark Holderman and Edward Henderson of the Technology Applications Assessment Team) working for NASA's Institute for Advanced Concepts (NIAC), the Nautilus-X was designed for interplanetary travel for a crew of six.

So, what happened to this revolutionary spacecraft? I spoke with Mark Holderman, a NASA rocket scientist, and he explained that a change in priorities and lack of funding ultimately canceled the project. A real shame.

So where does this leave us? There is limited research being conducted on artificial gravity, but before we take a look at this, it is worth mentioning the Centrifuge Accommodations Module (CAM). The CAM (◘ Fig. 9.3) was yet another missed

Fig. 9.2 The Non-Atmospheric Universal Transport Intended for Lengthy United States Exploration, aka Nautilus-X, was projected to cost $3.7 billion (compare this with $12 billion and counting for the Orion capsule... with no artificial gravity!). It was also projected to take only 5 years to complete (Orion, by comparison, has taken more than a decade and it still isn't ready!). Credit: NASA

opportunity in the field of artificial gravity research, as this module was supposed to have been part of the ISS. But guess what happened? You guessed it: budget cuts!

By now, you may be wondering why this artificial gravity nut is such a hard one to crack. Well, building a spacecraft that can spin is only half the task. The physiological responses to continuous exposure to artificial gravity are poorly understood. For example, what is the minimum level of gravity required to provide an effective countermeasure effect? What duration of rotation is required? How much of an effect will the Coriolis effect and cross-coupled accelerations have on astronauts? There is a long list of unknowns.

To give you an idea of the myriad challenges involved, let us take the idea of spinning the habitat around an axis as a way of creating artificial gravity. An astronaut standing on the rim of such a habitat that rotates at four revolutions per minute around an axis 56 meters long would experience about the same amount of gravity as on Earth (□ Fig. 9.4). Sounds like an elegant way to create artificial gravity, right? Well, yes and no. You see, when linear motion is created in a plane that is not parallel to the axis of rotation, *Coriolis force* is the result [2–4]. This

9

◘ Fig. 9.3   The Centrifuge Accommodations Module, which has been sitting in the car park at the Japanese Space Agency for more than a decade. Credit: NASA

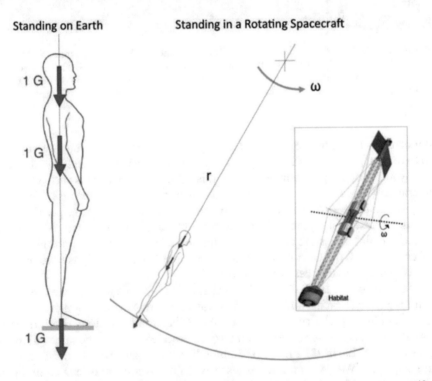

◘ Fig. 9.4   Artificial gravity. Rotating a spacecraft with a suitably long axis would create a centrifugal force of 1 G in the habitat. Credit: Clement et al.

Coriolis force combines with centrifugal force, with the result that a gravity vector is generated. The problem here is that the gravity vector is different in magnitude and direction.

How does this state exert its effect physiologically? Imagine our astronaut walking in their rotating environment again. This vector would add to the apparent weight of the astronaut moving in the direction of rotation but would subtract from the weight when the astronaut moves in the opposite direction of motion. And when the astronaut moved radially toward the center of rotation, the Coriolis force would be applied at right angles to the astronaut's motion in the direction of rotation [5–7]. This is bad enough, but if the astronaut displaced themselves at an angle to the spin axis, cross-coupled angular accelerations would create a problem for the neurovestibular system. Inside the neurovestibular system, you have semicircular canals that provide information about angular acceleration (i.e., pitch, roll, and yaw), and in a stationary environment, the only semicircular canals that are stimulated are those that correspond to the plane of your head's rotation. But in a rotating environment, that same head movement also stimulates the canals that are aligned with the plane of the rotating environment [8–10]. This is bad, because it creates the sensation of illusory self-motion, which in turn can cause motion sickness.

So, what can be done to avoid this illusory self-motion? The centrifugal force is dependent on rotation rate and radius, so it is these factors that must be manipulated. Of course, all these changes will affect the size, complexity, and cost of the spacecraft, but first it is necessary to determine what the optimum rate of rotation is and how long humans can be exposed to rotating environments without feeling adverse physiological effects. Many studies were conducted in the 1960s that tried to determine these factors (◘ Fig. 9.5), but many of these applied theoretical limits to rotation rates to determine the "comfort zone."

◘ **Fig. 9.5** Hypothetical comfort zone of artificial gravity based on studies conducted in the 1960s. Credit: T. W. Hall (from "Artificial Gravity in Theory and Practice." Conference Paper, July 2016: 46th International Conference on Environmental Systems (ICES), Vienna, Austria)

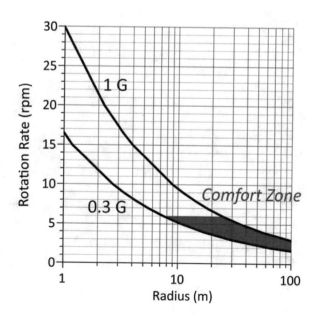

Artificial Gravity in Space

More recently, the focus has been on short-radius centrifugation. One reason for this is that building a rotating spacecraft presents a serious engineering and financial challenge. Another reason is that some scientists argue it may not be necessary to provide artificial gravity 24/7. Perhaps intermittent artificial gravity will do the trick? Who knows? One such human-rated centrifuge was flown on the STS-90 Neurolab mission in 1998 [3, 8]. More than 20 years later, this remains the first and only inflight evaluation of artificial gravity on astronauts. Incidentally, the Neurolab centrifuge generated between 0.5 and 1 G along the astronaut's longitudinal and transverse axis, and this was tolerated well. In this study, in those astronauts who were subjected to this G level for 20 minutes every other day, cardiovascular deconditioning was slightly reduced [3, 8].

Combining Exercise with Artificial Gravity

Back on the ground, scientists have experimented with contraptions that combine exercise with artificial gravity (◘ Fig. 9.6). The theory is that if astronauts exercise while they are being spun on a centrifuge, they won't have to spend as much time exercising on the regular suite of countermeasures. Call it multitasking.

During their time on these short-arm/short-radius centrifuges, test subjects lie supine with their head at the axis of rotation, and their feet pointed outward, which means the centrifugal force is along the longitudinal axis. In this configuration, the gravity gradient exerts an effect on hydrostatic pressure and therefore exerts an

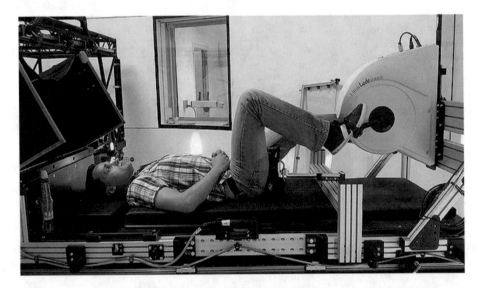

◘ **Fig. 9.6**   The next-generation short-arm centrifuge at DLR, Germany. Credit: DLR/ESA

effect on the cardiovascular system [11–13]. How much of an effect? Well, that's why they are conducting the studies. But what about that Coriolis force mentioned earlier? Will astronauts still suffer from that on the short-arm centrifuges? Recall that the Coriolis force is proportional to the linear velocity of the motion, the mass of the object being moved, and the rotation rate. This all means that the Coriolis force will still be a factor in short-arm centrifuges [12, 14].

But will it still be as much of a problem? Probably not. Look at ◙ Fig. 9.6; you can see how restricted the test subject is. This restricted position will help alleviate some of the problems encountered in the fictional rotating habitat configuration discussed earlier. So what is the right prescription of gravity and exposure time on board a short-arm centrifuge? What duration and frequency should be applied? How will undesirable side effects be countered? How much artificial gravity training will be required by crews? How should the G-loads be applied? What pharmacological interventions will be required to optimize crew health? What onset and offset rates should be applied? How effective are specific G-loads at reducing muscle atrophy? We don't know, hence the requirement for many, *many* more studies.

Optimizing Variables

The good news for NASA is that it has time on its side, because even if you are the most reckless of optimists, the agency's manned Mars mission won't be happening until at least the 2040s or 2050s. That gives researchers time to conduct studies into bedrest and intermittent centrifugation to determine how much bone and muscle atrophy is reduced and to develop all the aforementioned protocols. Having said that, you can conduct all the bedrest short-radius centrifuge studies in the world, but at the end of the day, you are still not replicating exactly what will happen in space.

In 2020, bedrest studies dedicated to solving the artificial gravity puzzle are few and very far between. There have been some studies that evaluated sensorimotor performance [10, 15] and a few investigations that have assessed the utility of partial gravity analogs [15], but most of these studies exposed their subjects to test durations of days and weeks, which is far short of the 12 months required for a return trip to Mars. So, let's take a look at some of those variables again: onset rate, offset rate, G-load, period of exposure to G-load, adaptation time following offset of G, and exercise. Now, think of just one permutation. We'll consider an onset rate of 0.2G/sec and have the subject spend 2 hours per day at 1-G while completing exercise at 75 percent of their maximal oxygen uptake with an offset rate of 0.2G/sec and an post-G adaptation time of 2 hours. Each subject will follow this protocol in a bedrest facility for 12 months. We'll call this a pilot study, so we'll select just 24 subjects, 12 male and 12 female. After 1 year, the scientists can crunch the numbers and start another protocol. Take another look at the number of variables and now think how many permutations of those variables are possible. Such a study would take...well, it would take a long time!

So there is a long way to go in the pursuit of artificial gravity. And while ground-based studies will help determine the potential for artificial gravity, ultimately these studies will need to be validated in space. In 2020, no human-rated centrifuge built to counteract physiological deconditioning has been flown in space since STS-90/Neurolab, and no such device is even on the drawing board. Hopefully this state of affairs changes soon, and perhaps with a concerted effort, we may see some form of artificial gravity by the end of the twenty-first century.

Hibernation

Hibernation, or *stasis* as it is occasionally referred, is a concept that is routinely used in science fiction movies (*Alien, Aliens, Avatar, Interstellar, Pandorum, Passengers,* to name just a few). If you think hibernation is a little far-fetched, consider an article published in the *British Medical Journal* over 100 years ago entitled "Human Hibernation" [16]. The article, reprinted in 2000, describes how Russian peasants survived famine by sleeping for up to 6 months of the year. The sleep was known as *lotska*. Here's an excerpt:

9

» A practice closely akin to hibernation is said to be general among Russian peasants in the Pskov Government, where food is scanty to a degree almost equivalent to chronic famine. Not having provisions enough to carry them through the whole year, they adopt the economical expedient of spending one half of it in sleep. This custom has existed among them from time immemorial.

At the first fall of snow the whole family gathers round the stove, lies down, ceases to wrestle with the problems of human existence, and quietly goes to sleep. Once a day every one wakes up to eat a piece of hard bread, of which an amount sufficient to last six months has providently been baked in the previous autumn. When the bread has been washed down with a draught of water, everyone goes to sleep again. The members of the family take it in turn to watch and keep the fire alight.

After six months of this reposeful existence the family wakes up, shakes itself, goes out to see if the grass is growing, and by-and-by sets to work at summer tasks. The country remains comparatively lively till the following winter, when again all signs of life disappear and all is silent, except we presume for the snores of the sleepers.

Animal Hibernation

In nature, hibernation is a time when arctic ground squirrels, polar bears, lemurs, and all the other hibernators of the animal world "sleep" through cold weather. But this sleep isn't like human sleep. In true hibernation, animals can be moved around and not know it, although you probably wouldn't want to test this theory with a polar bear! You see, hibernation is just one of the five forms of dormancy observed in hibernating animals. The other forms of dormancy are sleep, torpor, winter sleep, and summer sleep.

To prepare for their hibernative increment, animals tend to eat extra food in the autumn and pack on the pounds needed to survive the hibernation period. Having increased their waistline, hibernators search for a suitable place to hibernate – the *hibernaculum*. It is thought that some animals, such as arctic ground squirrels, enter hibernation, thanks to a "trigger molecule" that initiates hibernation. This molecule is termed the *hibernation induction trigger* (HIT), and its action is not completely understood, but it is possible a similar technique may be used to hibernate astronauts. Why would we do this? The obvious answer is to reduce demand on the life support system (◘ Table 9.1).

Once a hibernating animal enters hibernation, a number of things happen. In the ground squirrel (sidebar), respiratory rate drops from 200 breaths per minute to just 4–5 breaths per minute, and heart rate falls from 150 to 5 beats per minute. The fall in breathing rate and heart rate result in a reduction in metabolic rate, but this change doesn't stay the same throughout hibernation, because hibernating animals occasionally wake to eat, drink, and eliminate wastes. During these wakeful periods, body temperature and other physiological parameters return to normal levels. During hibernation, animals use 70 to 100 times less energy than when active (to maintain basic physiological function, these animals get energy in the form of adenosine triphosphate (ATP), which is produced in the mitochondria). Once the animal exits

◘ **Table 9.1** Effect of hibernation on life support system requirements

Life support area	Purpose	Effect of hibernation
Atmosphere management	Atmosphere control, temperature/humidity control, atmosphere regeneration, ventilation	Reduced heating and regeneration requirement
Water management	Provision of potable and hygienic water, recovery and processing wastewater	Reduced dramatically
Food production/ storage	Provision and production of food	Reduced dramatically
Waste management	Collection, storage, and processing of human waste and refuse	Reduced dramatically
Crew safety	Fire detection and suppression	Increased
Crew psychology	Maintenance of crew mental health	Reduced dramatically
Crew health	Bone demineralization and muscle atrophy	Augmented systems required

hibernation, the biochemistry and metabolism return to normal, although the animal, having woken after spending months asleep, may feel a little discombobulated!

Human Hibernation

So, it seems we have a reasonable understanding of how animals hibernate, but how would you induce an extended comatose state in a human? Although the procedure is currently barely beyond the science fiction arena, thanks to research efforts conducted by the European Space Agency (ESA) and SpaceWorks, this technology may one day be operational.

Let us focus on the work of ESA first. ESA scientists have suggested astronaut hibernation strategies may mirror those of the arctic ground squirrel, animals that undergo three defined periods of hibernation: entry, the hibernation period, and exit. We will now take a look at how astronauts might prepare for each of these stages.

Stage 1: Entry

First, the crew would be required to attain a high level of fitness to maximize their bodies' ability to deal with the hibernative period. Once en route, the crew would be connected to intravenous tubes, through which fluids and electrolytes would be administered to compensate for changes in blood composition. Then, the HIT would place the crew in a state of hibernation. The key to putting astronauts in a state of hibernation may lie in a synthetic, opioid-like compound called *DADLE*, or *Ala-(D) Leuenkephalin*, which, when injected into squirrels, can trigger hibernation during the summer.

Stage 2: Hypersleep Increment

During the hibernation phase, a suite of medical sensing facilities would monitor the state of the crew. In addition to checking that body temperature, heart rate, brain activity, and respiration stayed within normal boundaries, the medical equipment would also monitor blood pressure, blood glucose levels, and blood gases. Exactly what would happen to the astronauts as they hibernated isn't known. While humans have never needed to hibernate for protection against the elements (except those Russian peasants!), is it possible that we once had the biological mechanisms to regulate our metabolic activity and temperature for long periods of time? Do we still have those mechanisms and just not use them perhaps? Until scientists perform more studies, we won't know.

Stage 3: Exit

After their interplanetary trip, astronauts would be revived. This event would probably occur shortly before entering orbit due to the deleterious effects of having been in hibernation. But to what degree would hibernation have affected the crew? The truth is, no one knows. In the film *Pandorum*, hibernation leaves crewmembers with amnesia, although they recover later. Research has suggested the deep torpor associated with hibernation may be problematic for the brain, so having a handbook outlining the disorientation recovery procedures will probably be helpful (Appendix VI). The discombobulating effects of hibernation occur during torpor entry, when the body's temperature is gradually reduced. The cooling process results in reduced cortical power and profound differences in sleep architecture and memory consolidation. More worrying for those awakening from hibernation are the potentially deleterious effects upon spatial memory and operational conditioning. Imagine waking up from hibernation and not knowing where you are. Of course, until hibernation is performed on humans, we just won't know for sure. It is possible that humans will need only to follow a few instructions written in a Hypersleep Recovery Procedures (HRP) manual (Appendix VI) to fully recover, or perhaps more insidious effects may be the consequence.

How Squirrels Hibernate

Richardson's ground squirrels, also called gophers, hibernate for 4–9 months of the year, depending on age and gender. Each animal hibernates underground by itself in its own hibernaculum. The squirrels spend 85–92% of hibernation in the physiological state of torpor, during which time their body temperature is about the same as the temperature of the surrounding soil, and heart rate, respiration, and metabolism slow dramatically. In January, these squirrels spend 20 to 25 consecutive days in torpor, with their body temperature dropping as low as 0 °C. In between the periods of torpor, the squirrels rewarm to the normal mammalian body temperature of 37 °C. The revivals last less than 24 hours and consist of a 2–3-hour rewarming period, followed by 12–15 hours when the animal is warm but mostly inactive. Body temperature then slowly cools back down to ambient soil temperature, and the squirrel enters another period of torpor. Generally, the colder the soil, the colder the squirrel and the longer the period of torpor.

During hibernation, the squirrels metabolize fat reserves built up during their active season. Most of this fat is used during revival periods between during hibernation when the squirrel rapidly warms up and stays warm for several hours. Thus, arousals are metabolically expensive. Males usually end their hibernation about a week before they appear aboveground, while females end it the day before they appear aboveground.

Another way astronauts may be placed in a hibernative state is via a process known as *therapeutic hypothermia*. This procedure is being investigated by SpaceWorks, thanks to funding from NASA's Innovative Advanced Concepts Program. Therapeutic hypothermia places astronauts in an inactive, low-metabolic state, also known as torpor. Science fiction? Not at all. Therapeutic hypothermia is used in hospitals to place patients into sedated hypothermia following a serious injury

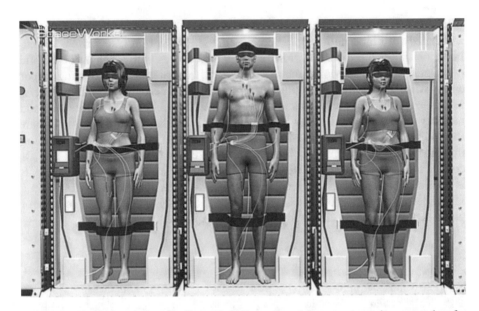

◘ Fig. 9.7 The habitat, designed by SpaceWorks Enterprises, Inc., can support six crewmembers for a duration of 180 days outbound, a 500-day surface stay, and 180 days inbound. To prevent muscle atrophy, neuromuscular electric stimulation is used (this is still a developing technology), and nutrients are delivered via central venous catheter in the chest (note: hibernation is not for the squeamish!). Credit: SpaceWorks/NASA

to allow them to heal while in a low metabolic state. This is achieved by lowering the body temperature by 5 °C. It doesn't sound like much, but when you do this, the body's metabolic rate is reduced by 50 to 70 percent. Of course, the astronauts will still need food, but that is accomplished via a technique known as *total parenteral nutrition* (TPN), which basically means a catheter delivers liquid nutrients directly into the bloodstream (◘ Fig. 9.7).

Monitoring

How will astronauts be monitored? Most likely, this will feature an AI system capable of monitoring medical parameters such as body temperature, electrocardiogram, heart rate, brain activity, gas exchange, blood pressure, and a whole host of other variables ranging from blood glucose and metabolite levels to gas analysis and clotting times. Data would be passed on to an AI hibernative agent – an avatar of sorts – and to mission control for analysis. The AI hibernative agent would act as a surrogate nurse for the duration of the voyage. It would be capable of interpreting medical information supplied by the monitoring system and acting on that information in a timely manner. If a problem developed, the agent might consult with mission control. If it was judged the contingency was serious and the communication delays too long, the agent would intervene autonomously.

This AI would probably be organized on two levels. The higher level would monitor fault detection, diagnosis, and planning, while the lower level would be responsible for perception, data acquisition, and dealing with messages from flight surgeons at mission control. The AI would probably be loaded with a database organized under six organ systems (cardiovascular, pulmonary, renal, hematological, neurological, and metabolic/endocrine). This repository would contain information on diseases/complications and treatment actions and plans. Because of the need to operate autonomously in the event of a medical emergency, the agent would be capable of three major reasoning components. The first of these would perform data analysis and interpretation, the second would perform diagnoses and therapy management, while the third would perform protocol-based treatment. A central monitoring computer (CMC) would be used as the core element of the sensor monitoring system. The CMC would gather data sent by medical sensors and log and update data gathered in the central database. Each astronaut would wear a sensor unit that would monitor and transmit their vital signs to the CMC.

Life Support

Thanks to the astronauts sleeping to their destination, LSS requirements could be scaled down. For example, food production and preparation facilities will be redundant, and hygiene facilities will obviously be scaled down and perhaps even eliminated altogether. One aspect of life support that will need to be considered is the inclusion of artificial gravity due to the prolonged immobility of the crew. It's possible the hibernaculum will actually be integrated into an artificial gravity facility to prevent the crew from losing too much bone and muscle.

If thoughts of long-duration space journeys and hibernation conjure up images of the opening scene of *Alien*, you're not alone. Although placing astronauts into hibernation would solve many problems during the deep space phases of an ECM, several issues remain unresolved. Scientists still need to develop a trigger (HIT) compound capable of inducing a state of hibernation and research the secondary effects of hibernation. For example, the effects of hibernation on memory, the metabolism, or the immune system are unknown. Other problems include the deleterious effects of zero gravity combined with the inactivity of hibernation, although this may be resolved by using some means of artificial gravity. Another challenge is how to induce the hibernation state, establish it, regulate it, and exit that state and how to administrate compounds to a hibernating human. Achieving and perfecting human hibernation will require expertise in and integration of pharmacology, genetic engineering, environmental control, medical monitoring, AI, radiation shielding, therapeutics, spacecraft engineering, *and* life support. Only when *all* these disciplines have been successfully integrated will stasis be capable of making long-haul spaceflight a little more comfortable.

At the moment, the level of inquiry really is just speculative, but while stasis may seem the stuff of science fiction, as so often happens, science fiction has a habit of becoming fact.

Bioprinting

》 There is no such thing as a science fiction. There is only science eventuality.
　　Professor Krummel. Chair of Department if Surgery, Stanford University, CA

Imagine the following scenario: Sometime in the near future, you have a three-dimensional scan of your body. While driving back from the medical clinic, you're involved in a head-on collision with another car. You lose your right ear and your left arm below the elbow. Worse, the missing body parts are so badly mangled they can't be stitched back. Fortunately, thanks to that 3D scan, your doctor selects the specs on your missing body parts and *prints* the body parts, which are identical to the ones you lost. You have surgery the same day to have the body parts reattached, and the doctor sends you on your way.

Biofabrication

Welcome to the world of bioprinting, aka *biofabrication*, a future of organ and body part replacement technologies produced by the latest breakthroughs in reconstructive medicine and bioprinting. Need a spare organ? A new liver perhaps? No problem. Just press "Print!" Now, printing a kidney or another human organ may sound like something out of a science fiction novel, but with the advancements in 3D printing technology, the idea is no longer so far-fetched. In fact, in 2019 a bioprinter was flown to the ISS, and in 2020, the astronauts started bioprinting heart tissue! Today, the technology is used to manufacture everything from prosthetic limbs to dental fixtures. One of the leaders in this biotech revolution is Organovo, a San Diego-based company that focuses on regenerative medicine. To print functional tissue, Organovo uses 3D printers, called *bioprinters* (◘ Fig. 9.8).

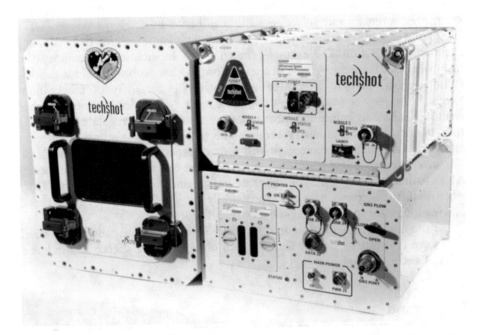

◘ **Fig. 9.8**   The bioprinter currently in use on board the ISS. Formally known as the BioFabrication Facility (BFF), the printer, manufactured by Techshot, is used to print cardiac tissue of various thicknesses. Credit: NASA

So where does bioprinting fit into the world of manned spaceflight? Well, manned spaceflight has seen significant changes over the past few decades, the most recent development being the manning of the ISS since 2000. And thanks to the success of the ISS, manned missions to the Moon and Mars are the logical next step. But these missions represent a huge step in technical, scientific, and medical complexity. The duration of Mars missions will raise several health issues, some of which may ultimately be limiting factors (see ◘ Table 9.2 for a list of some of these). Furthermore, given the impossibility of an abort, manned Mars missions will need to be self-sustainable. Given such a mission scenario, the integration of a technology such as biofabrication for regenerative medical support and organ reproduction is an essential enabling technology [17].

Biofabrication is a technology that uses biological raw materials, extracellular matrices, living cells, and tissues to construct something that is different from its components. The technology basically comprises six essential elements:

- A computer-aided design (CAD) drawing of the desired organ (think of this as a blueprint)
- Cells or cell-encapsulated *hydrogels* capable of natural self-assembly (this is referred to as *bioink* in the biofabrication world)
- A bioprinter for printing the bioink
- A biocartridge of the material to be deposited
- A bioprocessible biomimetic hydrogel to transfer material
- A container containing the resulting printed 3D tissue construct capable of post-conditioning (bioreactor)

Preprocessing

The first step is *preprocessing*. This requires a blueprint of the structure to be printed. Typically, this blueprint is generated using CAD, which also provides information about the cells' location. Once this information is generated, the digitized image is reconstructed using bio-imaging and/or image acquisition techniques. These techniques, such as magnetic resonance imaging (MRI), are used to capture the image in a way that provides detailed representations of the gross anatomy of organs. These techniques provide a good representation of the structure and cellular-level details such as tissue composition and distribution. If more detailed information of the tissue composition and the size and shape of the organ is needed, serial histological sections can be used to reconstruct 3D representations. Once the blueprint is designed, the organ can be printed and solidified [19].

Processing

This step, processing, utilizes devices designed to deliver and deposit material onto a substrate. Using the blueprint, the layer-by-layer placement of natural materials (typically cells) is achieved using a bioprinter. Before this stage occurs, bioink is prepared, and the resulting bioink droplets loaded into a biocartridge. Bioink is then sent through a syringe-like nozzle and deposited onto biopaper. The desired cell constructs are created by precise placement of the cells based on the blueprint. During the printing process, layers of cells are separated by a thin layer of biomimetic hydrogel to allow the resulting 3D tissue structure to fuse [20, 21].

9

◼ **Table 9.2** Probability occurrence of health issues during three mission scenarios [18][1]

Health challenge	Scenario 1 (180 days Moon mission)		Scenario 2 (1,000 days Mars mission)		Scenario 3 (500 days Mars mission)	
	On board spacecraft	On lunar surface	On board spacecraft	On Mars surface	On board spacecraft	On Mars surface
Respiratory infections	0.35	7.9	39.3	23	28.2	1.3
Cystitis	0.08	1.8	8.8	5.2	6.3	0.3
Skin infections	0.08	1.8	8.8	5.2	6.3	0.3
Cardiovascular disease	AR	0.02	0.1	0.06	0.07	0.003
Fracture of the skull	AR	0.004	0.02	0.01	0.01	AR
Spinal fracture	AR	0.004	0.02	0.01	0.01	AR
Upper limb fracture	AR	0.01	0.06	0.04	0.04	0.002
Lower limb fracture	AR	0.006	0.03	0.02	0.02	AR
Sprains and strains	0.006	0.13	0.7	0.4	0.5	0.03
Head injury	AR	0.004	0.02	0.01	0.01	AR
Open wounds	AR	0.02	0.1	0.06	0.07	0.004
Superficial injury	AR	0.02	0.1	0.06	0.07	0.004
Contusion	0.002	0.04	0.2	0.1	0.1	0.007
Crushing injury	AR	0.02	0.1	0.06	0.07	0.003
Burns	AR	0.02	0.1	0.06	0.07	0.003
Dental issues	AR	0.02	0.1	0.06	0.07	0.003

AR accepted risk, because the probability of occurrence is too low (<0.001); *BFO* blood forming organs
[1]Adapted from [18]
Table note: obviously, not all these health challenges can be solved with bioprinting, but most can be solved with either bioprinting or nanotechnology, which is discussed in the following section

Post-processing

After the processing stage, all you have is printed tissue and organ constructs. Because these aren't yet mature functional tissues, what is needed is a rapid process of self-assembly, maturation, and differentiation – the *post-processing* stage. Prior to post-processing, the constructs have the physical properties of a viscoelastic fluid, whereas actual organs usually have the physical properties of an elastic solid. So, for the constructs to become solid organs, they have to undergo accelerated tissue maturation, which can only be achieved in a wet environment using a special perfusion device – a *bioreactor* – that allows the cells to survive. Basically, what the bioreactor does is to ensure stable chemical and mechanical conditioning of the printed tissue and organ construct. To accurately bioprint organs, it is necessary to deposit the cells in a very precise manner. Today, this is achieved by using modified bioprinters/deposition devices that use robotically controlled syringe-like delivery tools [22, 23]. Once these specialized delivery tools have deposited the cells, the printed tissue constructs grow and develop in a wet environment, similar to the conditions found in the body. We'll go into a little more detail in the next section.

Bioprinting Tech

We'll start with an explanation of the most popular tissue engineering approach, a technique known as *solid scaffold-based biofabrication*. This technology was invented by Robert Langer and Joseph Vacanti at the Massachusetts Institute of Technology (MIT). In Langer and Vacanti's technology, a scaffold is a temporary supporting structure and is biodegradable. These scaffolds, which can be synthetic or naturally derived, are laid down in a top-down approach and have been used to create relatively simple tissue-engineered bladders. While this technology works for creating simple body structures, the seeding or injection of stem cells into a decellularized matrix – which is essentially what this technology does – probably won't work for complex organs such as the heart. That's because the technology uses animal-derived xenogeneic (meaning they are derived from a different species) scaffolds which, while suitable immunologically for a bladder of a bronchus, won't work for a liver or a lung. So, to get around this problem, scientists are investigating the use of scaffolds produced by using living human cells. These are allogeneic, meaning they are genetically different because they are derived from separate individuals of the same species and therefore work much better immunologically than animal-derived xenogeneic scaffolds [24].

Cell Sheet Technology

Another popular approach in tissue engineering is *cell sheet technology*, a biofabrication technique that may be applied to the construction of heart valves. Cell sheet technology, as its name suggests, comprises a solid scaffold-free self-assembly process that utilizes stacked or rolled layers of engineered tissue fused to form thicker constructs. This technology has already been used to build the first completely biological tissue-engineered vascular graft.

But what about printing organs? For organ printing, scientists must utilize more complex technology because, well, organs are complicated parts of your body. One way to print organs is to use *directed tissue self-assembly*, which works by employing self-assembling tissue spheroids as building blocks. When these building blocks are placed close together, fusion occurs, a process that happens to be ubiquitous during embryonic development. And because this process mirrors what happens biologically, it is said to be *biomimetic*. Already, this technology has been used to bioprint a branched vascular tree.

While facing serious technical challenges, step by step, this organ printing technology will eventually be used to build 3D vascularized functional human organs (and ultimately, these organs will be printed in space – and on the surface of Mars!). In fact, some bioprinting techniques are designed specifically with certain organs in mind. Take *centrifugal casting*. This technology allows scientists to fabricate tubular scaffolds with high cell density in a porous scaffold, and while this technology isn't sufficiently versatile to biofabricate all the organs in the body, there are plenty of tubular organs that can be made [25].

Electrospinning

Another fast-developing biofabrication technology is *electrospinning*, a process that interfaces nanotechnology and tissue engineering. Originally used on synthetic and natural polymers, the technology can be applied to nanoscale tissue-engineered scaffolds. It's an important technology in biofabrication because it allows scientists to overcome some of the problems encountered when building structures on a microscopic scale. One of these challenges is fabricating three-dimensional scaffolds to support cellular in-growth and proliferation. Once constructed, these neotissues must be adapted by the body to carry out a physiological function. Here, the process becomes difficult, because the cells on the surfaces of the scaffolds are sensitive to topography. This is where electrospinning comes into its own, because the technology can be altered to change the surface topography of the fibers themselves or the topography of the web of spun fibers.

Like most cutting-edge technologies, there are still some bugs. For example, one of the problems of electrospinning is cell seeding; while dense biomimetic nano-structuralized matrices are great for getting cells to stick, they aren't so good when it comes to allowing cells to pass through the matrices. One way to overcome this is to use *cryospinning*, which allows scientists to create custom-sized holes in electro-spun matrices, but then there is the problem of compromising the biomechanical properties of the scaffold.

Game-Changing Technology

Scientists already know how to bioprint tissues, and the capability of bioprinting organs doesn't seem that far over the horizon. But what about other body parts such as bones or, say, a trachea? Here, progress is being made. Consider the case of Claudia Castillo, 30, a Colombian woman who in 2008 became the world's first recipient of windpipe tissue constructed from a combination of donated tissue and her own cells. Ms. Castillo had suffered a collapse of the tracheal branch of her windpipe leading to her left lung following a severe tuberculosis infection. As she was barely able to breathe, doctors decided to attempt trachea reconstruction by taking a 7-cm section of the trachea from a deceased donor. Researchers at the University of Padua, Italy, used detergent and enzymes to purge the donated trachea of all the donor's cells until all that was left was a solid scaffold of connective tissue. Meanwhile, a team from Bristol in the United Kingdom took the stem cells from Ms. Castillo's bone marrow and coaxed them into developing into the cartilage cells that normally coat windpipes. Then, Ms. Castillo's cells were coated onto the donated tracheal scaffold in a special bioreactor, after which the biologically printed trachea was ready to be transported to Barcelona, where surgeon Paolo Macchiarini was waiting to replace Ms. Castillo's damaged trachea with the newly constructed tissue. The entire procedure cost $21,000.

So, scientists can bioprint a trachea, and they are confident they will be able to bioprint a kidney in the near future. But what about more complex organs like livers, and how will scientists integrate these body parts into the body so the tissues are kept alive and the organs function as they should? Let's take the first problem. It will be challenging to bioprint complex organs, but researchers are confident because while every organ type and tissue structure has its own complicated internal architecture, researchers believe there are basic cell patterns that, once fully understood, can be readily duplicated by bioprinting. When it comes to the challenge of integrating all these organs and body parts, biomedical engineers say they haven't yet mastered ways to print the microscopic networks of capillaries that run between layers of cells to keep normal tissue alive. The challenge of figuring out how to feed tissues is a big one, because there must be a "bridge" from the organ to the new host; a new organ has arteries and veins which must be hooked up to the patient's corresponding arteries and veins.

At the current level of bioprinting technology, a regular transplant will prolong life for much longer than anything created in a lab or a spacecraft. That's because lab-generated/printed organs can't really be considered as organs, since these organs are much less sophisticated than the ones naturally found in the body. For example, a regular liver is comprised of several dozen types of cells, each of which performs a specific function. But the livers researchers are making in the lab only have a few cell types in them, which are nowhere near the complexity naturally found in the body. So, the only use for these *organoids*, as researchers have dubbed them, is short-term prolongation of life. Another problem yet to be resolved is sustaining the organ once it is in the body. One of the ways to ensure the organ isn't rejected is to have the patient's immune cells migrate back in, as long as the organ isn't initially rejected [26, 27]. Problem is, researchers don't fully understand these cells, never mind being able to bioprint them. But work is underway, and gradually researchers are learning more and more about how to grow human cells outside of the human body. There is also a lot of support for this area from pharmaceutical companies, and with enough financial support, researchers will be able to break through a lot of these barriers.

Nanotech

In the discussion on the subject of hibernation, we mentioned the challenge of maintaining muscle mass in astronauts while they slept. Another problem is mitigating the problem of bone demineralization. And then there is the radiation issue and all those potential health issues that may be encountered (see ◘ Table 9.2). How can these challenges be met? Nanotechnology may have the answer.

Consider the following scenario. Built using nanomaterials called *dendrimers*, molecule-sized sensors could be injected into our hibernating crew to warn of radiation health impacts and repair damage caused by radiation. The drug-delivery capsules are tiny, measuring only a few hundred nanometers – smaller than a bacterium. An injection with a hypodermic needle could release millions of these capsules into the astronaut's bloodstream. Once there, nanoparticles could use the body's cellular signaling system to hunt down and repair radiation-damaged cells. How? Well, the trillions of cells in the human body identify themselves and communicate with each other using molecules embedded in their membranes. These molecules act as chemical "flags" for communicating with other cells, and when cells are damaged by radiation, they produce markers and place them on their outer surfaces. Basically, it's a system whereby cells talk to each other and say "Hey, I'm damaged." What nanoscientists could do is implant molecules on the outer surface of the nanoparticles that bind to the markers and then program the nanoparticles to search for radiation-damaged cells. Once the nanoparticles found the damaged cells, they could assess the extent of the injury. If the cell was very badly damaged, the nanoparticles could enter the cell and program it for destruction. If the damage was judged repairable, the nanoparticles would release DNA repair enzymes and fix the cell (◘ Figs. 9.9, 9.10, and 9.11).

Vasculoid

While the nanosensors in the bloodstream could help astronauts monitor radiation damage and even repair it, there's still a chance that even a system as versatile as this could be overwhelmed by a solar flare event. Such an event could put vital blood-making cells in jeopardy, and without a fresh supply of red and white blood

◘ **Fig. 9.9** Cell rover. These nanobots patrol the circulatory system, searching for breaches. © E-spaces and Robert A. Freitas Jr., ▶ 3danimation.e-spaces.com and ▶ www.rfreitas.com, and Philippe van Nedervelde

Fig. 9.10 Dendrimer complex docking on cellular folate receptors. © E-spaces and Robert A. Freitas Jr., ▶ 3danimation.e-spaces.com and ▶ www.rfreitas.com, and Philippe van Nedervelde

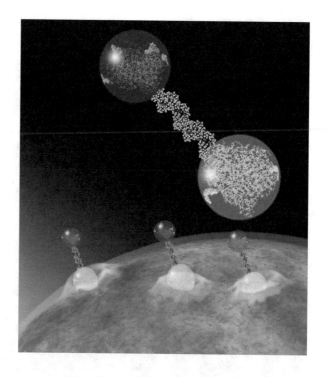

cells, our crew would quickly become anemic, their immune system would collapse, and without medical attention, they would die. So why not simply replace the blood with a more rugged system? Well, that's exactly what nanoscientists Robert Freitas and Christopher Phoenix propose doing. Freitas and Phoenix's concept involves exchanging a person's blood with 500 trillion oxygen and nutrient-carrying nanobots [28–30]. The system is called the *vasculoid* (a vascular-like machine) and is designed to duplicate every function of blood, albeit more efficiently. A key element of the vasculoid is the *respirocyte*. Each respirocyte is constructed of 18 billion precisely arranged atoms and has an on-board computer, power plant, and molecular pumps capable of transporting oxygen and CO_2 molecules. Not only is it capable of duplicating all thermal and biochemical transport functions of blood, but it is also able to perform these functions hundreds of times more efficiently than biological blood. In essence, the vasculoid is nothing short of a mechanically engineered redesign of the human circulatory system. Despite the complexity of the system (it comprises 500 trillion independently cooperating nanobots), it weighs only 2 kg and is powered by nothing more than glucose and oxygen.

The key structural element of the vasculoid is a 2D vascular surface-conforming array of 150 trillion square sapphiroids. The sapphiroids are self-contained superthin nanorobots that cover the entire surface of all blood vessels in the body to one plate thickness. Of the 150 trillion plates, 24 trillion are molecule-conveying docking bays where tankers containing molecules for distribution can dock and load/ unload their cargo. Another feature of the array is the cellulock, of which there are 32.6 billion. At the cellulocks, boxcars carrying biological cells dock and load/

9

◘ **Fig. 9.11**   Respirocyte in a blood vessel surrounded by red blood cells. The respirocyte is a nano-bot capable of duplicating all thermal and biochemical transport functions of blood. © E-spaces and Robert A. Freitas Jr., ▶ 3danimation.e-spaces.com and ▶ www.rfreitas.com, and Philippe van Nedervelde

unload their cargo. The remaining 125 trillion plates are reserved for special equipment and other applications. All the plates have watertight mechanical interfaces comprising metamorphic bumpers along the perimeter of each plate, which allow the bumper to expand and contract in area. It's a feature that permits the system to flex in response to body movements.

A discussion of all the vasculoid components is beyond the scope of this chapter, so we'll focus on how the vasculoid could help astronauts to combat radiation sickness. Remember, ionizing radiation causes atoms and molecules to become ionized or excited. These excitations and ionizations can produce free radicals and damage molecules that regulate vital cell processes. Although regular cells can repair some cell damage, if the cells are too badly damaged, they can't be replaced quickly enough, and tissues fail to function. If this were to happen in a vasculoid-installed astronaut, the vasculoid would detect the damage and deploy vasculo-cytes. Vasculocytes are nanobots equipped with ambulatory appendages, manipulator arms, repair tools, on-board computers, communications, and power supplies. They patrol the vasculoid continuously, searching for maintenance and repair tasks such as plugging internal leaks and cleaning spills. In the event of radiation damage, these nanorobots would search out affected cells using molecular markers before destroying these cells [29, 30]. Sounds like a pretty neat repair mechanism, but how would this device be installed?

Installation

Installation would be a complex process that would begin with exsanguination and finish with a process known as vascular plating. After being sedated, the astronaut's natural circulatory fluids would be removed and replaced with installation fluids. This step would be followed by mechanical vascular plating, defluidization, and finally activation of the vasculoid, rewarming the astronaut. From start to finish, the installation process would take about 6 hours. What follows is the step-by-step process as it may occur in the near future.

Preparation

Preparation would begin 24 hours before installation, when the astronaut would receive an injection of 70 billion vascular repair nanorobots that would clean out any fatty streaks, plaque deposits, lesions, and vascular wall tumors. After completing their tasks, the repair devices would be exfused and the results downloaded to a computer. This information would be used by the surgeon to prepare a map of the astronaut's vascular tree to improve efficiency during plating and plate initialization. Once this step was complete, the astronaut would be sedated, cannulated, and hooked up to a heart–lung machine. Heparin and streptokinase would be injected to prevent clotting, after which the surgeon would administer various agents to aid the installation process.

Exsanguination

After the astronaut had been anesthetized, their entire blood volume would be replaced with a suspension of respirocytes, a mixture of electrolytes, and other components normally found in blood substitutes. The respirocyte fleet would provide oxygen and carbon dioxide transport equivalent to the entire human red blood cell (RBC) mass for 3 hours after the cessation of respiration. Once the blood volume had been exchanged, the astronaut's core temperature would be reduced from 37 °C to just 7–17 °C, after which the astronaut would be ready for the intravenous deployment of vasculoid components.

Intravenous Deployment

First, the respirocyte suspension would be replaced by a new suspension containing 1% fully charged respirocytes and 10% cargo-bearing vasculocytes, creating a mixture whose viscosity and flow characteristics approximate to human blood. Each vasculocyte would drift in the flow until it encountered a vessel wall, which would activate it, causing it to release its cargo. Once its cargo plate was in place, the vasculocyte would release back into the fluid, power down, and be exfused from the body. After positioning and subsystem validation, each plate would inflate fluid-tight metamorphic bumpers along its contact perimeter with its neighbors, which would lock their bumpers firmly together with reversible fasteners embedded in the bumpers. After about an hour, the structure of the vasculoid would be almost complete, and all the major components would have been tested. The astronaut would now be ready for defluidization.

Defluidization

During defluidization, a monolayer of nanorobotic plates would form a chemically inert, flexible sapphire liner on the vascular tree's interior surface, and vasculoinfusant fluid would be purged from the body by injecting 6 liters of oxygenated acetone to rinse the vascular tree. Once the system had been rinsed, the process of plate initialization would begin.

Plate Initialization

With 200 billion vasculocytes and 150 trillion plates to initialize, each vasculocyte would need to contact and initialize 750 plates. This stage would be followed by the installation of storage vesicles that contain reserves of mobile and cargo-carrying nanodevices and other auxiliary nanodevices. The astronaut would be rewarmed, the catheters would be removed, and the vascular breaches would be sealed. At this stage, the vasculoid would be operational, and essential metabolic and immunological systems would have returned to normal.

To many people, the installation of such a device into an astronaut for the purpose of protecting them from radiation damage may represent an extreme medical intervention. However, current knowledge of nanomechanical systems suggests such a device would not violate known physical, engineering, or medical principles [29, 30]. If in fact the vasculoid becomes a reality, it may represent a significant outpost not only in biological evolution but also in humankind's quest to extend the frontier beyond orbit.

Key Terms
- BioFabrication Facility (BFF)
- Central Monitoring Computer (CMC)
- Centrifuge Accommodations Module (CAM)
- Computer-Aided Design (CAD)
- Hibernation Induction Trigger (HIT)
- Hypersleep Recovery Procedures (HRP)
- Magnetic Resonance Imaging (MRI)
- Non-Atmospheric Universal Transport Intended for Lengthy United States Exploration (NAUTILUS-X)
- Red Blood Cell (RBC)
- Total Parenteral Nutrition (TPN)

❓ Review Questions
1. Describe three challenges in developing artificial gravity.
2. What is meant by the term *Coriolis force*?
3. What is meant by HIT?
4. Describe four LSS advantages of having a hibernating crew.

5. How do animals hibernate?
6. List the six essential elements of bioprinting technology.
7. What is cell sheet technology?
8. Describe the process of installing a vasculoid.

References

1. Von Braun, W. (1953). The baby space station: First step in the conquest of space. *Collier's Magazine, 27*, 33–35, 38, 40.
2. Antonutto, G., Linnarsson, D., & di Prampero, P. E. (1993). On-Earth evaluation of neurovestibular tolerance to centrifuge simulated artificial gravity in humans. *Physiologist, 36*, S85–S87.
3. Arrott, A. P., Young, L. R., & Merfeld, D. M. (1990). Perception of linear acceleration in weightlessness. *Aviation, Space, and Environmental Medicine, 61*, 319–326.
4. Benson, A. J., Guedry, F. E., Parker, D. E., & Reschke, M. F. (1997). Microgravity vestibular investigations: Perception of self-orientation and self-motion. *Journal of Vestibular Research, 7*, 453–457.
5. Clément, G., Moore, S., Raphan, T., & Cohen, B. (2001). Perception of tilt (somatogravic illusion) in response to sustained linear acceleration during space flight. *Experimental Brain Research, 138*, 410–418.
6. de Winkel, K., Clément, G., Werkhoven, P., & Groen, E. (2012). Human threshold for gravity perception. *Neuroscience Letters, 529*, 7–11.
7. Lackner, J. R., & DiZio, P. (2000). Human orientation and movement control in weightless and artificial gravity environments. *Experimental Brain Research, 130*, 2–26.
8. Benson, A. J., Kass, J. R., & Vogel, H. (1986). European vestibular experiments on the Spacelab-1 mission: 4. Thresholds of perception of whole-body linear oscillation. *Experimental Brain Research, 64*, 264–271.
9. Guedry, F. E., & Benson, A. J. (1978). Coriolis cross-coupling effects: Disorienting and nauseogenic or not. *Aviation, Space, and Environmental Medicine, 49*, 29–35.
10. Jarchow, T., & Young, L. R. (2010). Neurovestibular effects of bed rest and centrifugation. *Journal of Vestibular Research, 20*, 45–51.
11. Caiozzo, V. J., Rose-Gottron, C., Baldwin, K. M., Cooper, D., Adams, G., & Hicks, J. (2004). Hemodynamic and metabolic responses to hypergravity on a human-powered centrifuge. *Aviation, Space, and Environmental Medicine, 75*, 101–108.
12. Clément, G., Deliere, Q., & Migeotte, P. F. (2014). Perception of verticality and cardiovascular responses during short-radius centrifugation. *Journal of Vestibular Research, 24*, 1–8.
13. Greenleaf, J. E., Gundo, D. P., Watenpaugh, D. E., Mulenburg, G. M., McKenzie, M. A., Looft-Wilson, R., et al. (1996). Cycle-powered short radius (1.9 M) centrifuge: exercise vs. passive acceleration. *Journal of Gravitational Physiology, 3*, 61–62.
14. Clément, G., & Pavy-Le Traon, A. (2004). Centrifugation as a countermeasure during actual and simulated microgravity: A review. *European Journal of Applied Physiology, 92*, 235–248.
15. Wu, R. H. (1999). *Human readaptation to normal gravity following short-term simulated Martian gravity exposure and the effectiveness of countermeasures* (Master of Sciences Thesis). Cambridge, MA: Massachusetts Institute of Technology.
16. One hundred years ago: Human hibernation. *British Medical Journal, 320*(7244), 1245.
17. Ghidini, T., Pambaguian, L., & Blair, S. (2015). Joining the third industrial revolution: 3D printing for space. *Esa Bulletin, 163*, 24–33.
18. Horneck, G., Facius, R., Reichert M., Rettberg, P., Seboldt, W., Manzey, D., Comet, B., Maillet, A., Preiss, H., Schauer, L., Dussap, C. G., Poughon, L., Belyavin, A., Reitz, G., Baumstark-Khan, C., & Gerzer, R. (2003). HUMEX, a study on the survivability and adaptation of humans to long-duration exploratory missions, part I: Lunar missions. *Advances in Space Research, 31*(11), 2389–2401.

19. Jakab, K., Norotte, C., Marga, F., et al. (2010). Tissue engineering by self-assembly and bio-printing of living cells. *Biofabrication, 2,* 022001.
20. Boland, T., Xu, T., Damon, B., et al. (2006). Application of inkjet printing to tissue engineering. *Biotechnology Journal, 1,* 910–917.
21. Chan, B. P., & Leong, K. W. (2008). Scaffolding in tissue engineering: General approaches and tissue-specific considerations. *European Spine Journal, 17,* 467–479.
22. Ringeisen, B. R., Othon, C. M., Barron, J. A., et al. (2006). Jet-based methods to print living cells. *Biotechnology Journal, 1,* 930–948.
23. Mironov, V., Visconti, R. P., Kasyanov, V., et al. (2009). Organ printing: Tissue spheroids as building blocks. *Biomaterials, 30,* 2164–2174.
24. Saunders, R. E., & Derby, B. (2014). Inkjet printing biomaterials for tissue engineering: Bioprinting. *International Materials Reviews, 59,* 430–448.
25. Smith, C. M., Stone, A. L., Parkhill, R. L., et al. (2004). Three-dimensional bioassembly tool for generating viable tissue-engineered constructs. *Tissue Engineering, 10,* 1566–1576.
26. Gudapati, H., Dey, M., & Ozbolat, I. (2016). A comprehensive review on droplet-based bioprinting: Past, present and future. *Biomaterials, 102,* 20–42.
27. Murphy, S. V., & Atala, A. (2014). 3D bioprinting of tissues and organs. *Nature Biotechnology, 32,* 773–785.
28. Freitas, R. A., Jr. (1996, October). Respirocytes: High performance artificial nanotechnology red blood cells. *NanoTechnology Magazine, 2*(1), 8–13.
29. Freitas, R. A., Jr. (1998). Exploratory design in medical nanotechnology: A mechanical artificial red cell. *Artificial Cells, Blood Substitutes, and Immobilization Biotechnology, 26,* 411–430.
30. Freitas, R. A., Jr. (2001, October–December). Robots in the bloodstream: The promise of nano-medicine. *Pathways, The Novartis Journal, 2,* 36–41.

Suggested Reading

Seedhouse, E. (2011). *Trailblazing medicine.* New York: Springer-Praxis.
Seedhouse, E. (2012). *Interplanetary outpost.* Dordrecht: Springer-Praxis.
Seedhouse, E. (2014). *Beyond human.* Heidelberg: Springer Nature.

Supplementary Information

© Springer Nature Switzerland AG 2020
E. Seedhouse, *Life Support Systems for Humans in Space*,
https://doi.org/10.1007/978-3-030-52859-1

Appendix Ia: Ionizing Radiation

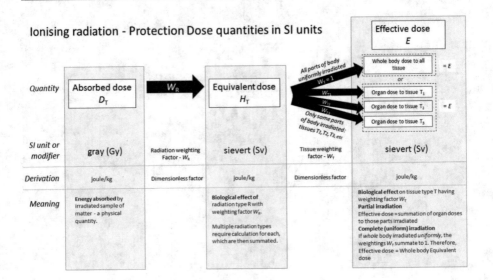

Ionising radiation - Protection Dose quantities in SI units

Quantity	Absorbed dose D_T	W_R	Equivalent dose H_T		Effective dose E
				All parts of body uniformly irradiated $W_T = 1$	Whole body dose to all tissue = E
					or
				W_{T1} W_{T2} W_{T3} Only some parts of body irradiated: tissues T_1, T_2, T_3, etc	Organ dose to tissue T_1
					Organ dose to tissue T_2 = E
					Organ dose to tissue T_3
SI unit or modifier	gray (Gy)	Radiation weighting Factor - W_R	sievert (Sv)	Tissue weighting factor - W_T	sievert (Sv)
Derivation	joule/kg	Dimensionless factor	joule/kg	Dimensionless factor	joule/kg
Meaning	**Energy absorbed** by irradiated sample of matter - a physical quantity.		**Biological effect of** radiation type R with weighting factor W_R. Multiple radiation types require calculation for each, which are then summated.		**Biological effect** on tissue type T having weighting factor W_T **Partial irradiation** Effective dose = summation of organ doses to those parts irradiated **Complete (uniform) irradiation** If *whole* body irradiated *uniformly*, the weightings W_T summate to 1. Therefore, Effective dose = Whole body Equivalent dose

Appendix Ib: Radiological Protection

Dose quantities in SI units for external radiological protection

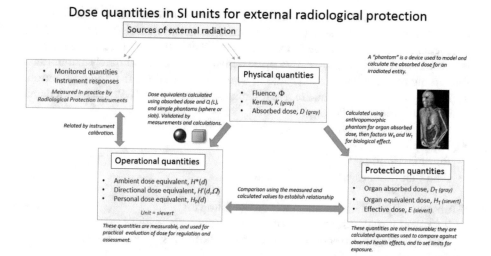

Sources of external radiation

- Monitored quantities
- Instrument responses

Measured in practice by Radiological Protection Instruments

Related by instrument calibration.

Dose equivalents calculated using absorbed dose and Q (L), and simple phantoms (sphere or slab). Validated by measurements and calculations.

Physical quantities

- Fluence, Φ
- Kerma, K (gray)
- Absorbed dose, D (gray)

A "phantom" is a device used to model and calculate the absorbed dose for an irradiated entity.

Calculated using anthropomorphic phantom for organ absorbed dose, then factors W_R and W_T for biological effect.

Operational quantities

- Ambient dose equivalent, $H^*(d)$
- Directional dose equivalent, $H'(d,\Omega)$
- Personal dose equivalent, $H_P(d)$

Unit = sievert

These quantities are measurable, and used for practical evaluation of dose for regulation and assessment.

Comparison using the measured and calculated values to establish relationship

Protection quantities

- Organ absorbed dose, D_T (gray)
- Organ equivalent dose, H_T (sievert)
- Effective dose, E (sievert)

These quantities are not measurable; they are calculated quantities used to compare against observed health effects, and to set limits for exposure.

Appendix II: Intracranial Hypertension

Intracranial hypertension is a disorder, the primary feature of which is persistently elevated intracranial pressure (ICP). The most pertinent neurologic sign of ICP is papilledema, which is discussed in ▶ Chapter 4. Papilledema is a red flag condition because it may lead to progressive optic degeneration and ultimately blindness – that's a big problem if you're en route to Mars or some other far-flung destination. Typical signs and symptoms are usually related to increased ICP and papilledema and may include the following:

— Headaches, often varying in type, location, and frequency
— Diplopia
— Pulsatile tinnitus
— Pain that radiates, usually in the arms

Symptoms of papilledema may include the following:
— Transient visual obscurations, often uniformly orthostatic
— Gradual loss of peripheral vision in one or both eyes
— Blurring and distortion of central vision
— Sudden visual loss

The most significant physical finding is bilateral disc edema secondary to the increased ICP. In more severe cases, edema and diminished central vision may be present. Visual function tests for diagnosing and monitoring patients with intracranial hypertension are similar to those used to assess VIIP and include:
— Ophthalmoscopy
— Visual field assessment
— Ocular motility examination
— MRI of the brain

The goal of managing the condition is to preserve optic nerve function while managing increased ICP, which is why common pharmacologic therapy may include acetazolamide, which is the most effective agent for lowering ICP.

For your reference, ◘ Table A.1 outlines the current NASA guidelines for ensuring ophthalmic health in the astronaut corps and ◘ Table A.2 provides a summary of the vision changes observed in seven astronauts shortly after the VIIP syndrome was reported.

For a more comprehensive assessment of this problem, the reader is referred to the International Space University's textbook on the subject: Microgravity and Vision Impairments in Astronauts by Erik Seedhouse, published by Springer.

◻ **Table A.1** Clinical practice guideline classifications of 36 long-duration US crewmembers

CPG class	Definition	# of affected astronauts
Non-cases	<0.50 diopter cycloplegic refractive change No evidence of papilledema, nerve sheath distension, choroidal folds, globe flattening, scotoma, or cotton wool spots compared to baseline	2
1	Repeat OCT and visual acuity in 6 weeks ≥0.50 diopter cycloplegic refractive change and/or cotton wool spot No evidence of papilledema, nerve sheath distension, choroidal folds, globe flattening, scotoma, or cotton wool spots compared to baseline CSF opening pressure (if measured) ≤25 cm H_2O	2
2	Repeat OCT, cycloplegic refraction, fundus exam, and threshold visual field every 4–6 weeks × 6 months, repeat MRI in 6 months ≥0.50 diopter cycloplegic refractive change or cotton wool spot Choroidal folds and/or optic nerve sheath distension and/or globe flattening and/or scotoma No evidence of papilledema CSF opening pressure (if measured) ≤25 cm H_2O	8
3	Repeat OCT, cycloplegic refraction, fundus exam, and threshold visual field every 4–6 weeks × 6 months, repeat MRI in 6 months ≥0.50 diopter cycloplegic refractive change or cotton wool spot Optic nerve sheath distension and/or globe flattening and/or choroidal folds and/or scotoma Papilledema of Grades 0–2 CSF opening pressure ≤25 cm H_2O	1
4	Institute treatment protocol as per CPG ≥0.50 diopter cycloplegic refractive change or cotton wool spot Optic nerve sheath distension and/or globe flattening and/or choroidal folds and/or scotoma Papilledema of Grade 2 or above Presenting symptoms of new headache, pulsatile tinnitus, and/or transient visual obscurations CSF opening pressure >25 cm H_2O	4
Unclassified	Too little evidence at present for definitive classification Early ISS flyers with no or limited testing	19

■ **Table A.2**   Summary of ophthalmic changes from seven affected long-duration crewmembers

Ophthalmic condition	Total affected
Optic nerve sheath distension	6/7 (86%)
Nerve fiber layer thickening	6/7 (86%)
Optic disc edema	5/7 (71%)
Posterior globe flattening	5/7 (71%)
Hyperopic shift in one the eye or both eyes by ≥ +0.50 diopters	5/7 (71%)
Choroidal folds	4/7 (57%)
Elevated postflight CSF pressure (indicative of increased ICP)	4/7 (57%)
Cotton wool spots	3/7 (43%)
Decreased intraocular pressure (IOP) postflight	3/7 (43%)
Tortuous optic nerve	2/7 (29%)

Appendix III: Psychology of Survival

For those who are interested, here are some more stories of survival. Read these and then ask yourself if expensive analogs are really needed to test human resilience. The truth is, anyone with even a rudimentary knowledge or experience of exploration knows that humans are a hardy bunch, as evidenced by these stories and many more like them.

Douglas Mawson

Even today, with all the advanced technology available, a journey on foot across Antarctica is one of the toughest tests imaginable (perhaps the Mars500 scientists should have conducted their research there), but a hundred years ago, it was worse. Much worse. Back then, polar explorers wore wool clothing that absorbed snow, and high-energy food was an unappetizing mix of fats called pemmican. But worst of all was the brutally cold environment. Apsley Cherry-Garrard, who sailed with Captain Scott's doomed South Pole expedition, reported that his teeth, after having been exposed to −60 °C, had "split to pieces." Cherry-Garrard wrote an account of his adventures in an aptly titled book *The Worst Journey in the World*. But even Cherry-Garrard's trek, made in total darkness in the depths of the Southern winter, wasn't as horrific as the harrowing march faced by Australian explorer Douglas Mawson. Mawson's journey has gone down in the annals of polar exploration as probably the most terrible ever undertaken in Antarctica.

In 1912, Mawson was 30 years old and acclaimed as one of the best geologists of his generation. He had declined the chance to join Scott's South Pole expedition so he could lead the Australasian Antarctic Expedition, the purpose of which was to explore and map some of the more remote areas of the Antarctic. An Antarctic veteran and gifted organizer, Mawson was as tough as they came, which, as events transpired, was a good thing. Mawson's team anchored in Commonwealth Bay in January 1912. Over the next few months, wind speeds averaged 80 kilometers per hour and sometimes exceeded 320 kmh! Blizzards were incessant and unrelenting. Mawson split his expedition into four groups, one to man base camp and three to conduct scientific work. He nominated himself to lead the Far Eastern Shore Party, a three-man team that was tasked with surveying glaciers hundreds of kilometers from base. It was a risky endeavor, because Mawson and his men would have the farthest to travel and the heaviest loads to carry.

Mawson selected Lieutenant Belgrave Ninnis, a British army officer, and Ninnis's friend Xavier Mertz, a Swiss lawyer, to join him. The explorers took three sledges, pulled by 16 huskies and loaded with 790 kilograms of food, survival gear, and scientific instruments. To begin with, progress was good, with Mawson's party traveling 480 kilometers in just 5 weeks. But after a series of near disasters, the men began to feel their luck was changing. Ninnis almost plunged into concealed cracks in the ice and Mawson suffered from a split lip that sent blinding pain shooting across the left side of his face. Worse was to come.

At noon on 14 December 1912, Mawson stopped to shoot the sun and determine their position. He was standing on the runners of his sledge, completing his calculations, when he became aware that Mertz, who was skiing ahead of the sledges, had raised one ski pole in the air to signal a crevasse. Mawson called to warn to Ninnis before returning to his work. A few minutes later, he noticed Mertz had stopped again and was looking back in alarm. Ninnis and his sledge and dogs had vanished. Mawson and Mertz hurried back to where they had crossed the crevasse and discovered a yawning chasm. Below, the walls plunged into darkness. Mawson called Ninnis's name for 5 hours, but there was no response. With practically all their food gone, all that remained was their sleeping bags and enough food to last a week and a half. They began their return to base, killing and eating the remaining dogs as they went, each night's rations less palatable than the last. Inevitably, the two men's physical condition deteriorated rapidly:

> » Starvation, combined with superficial frostbite, alternating with the damp conditions in the sleeping-bags, had by this time resulted in a wholesale peeling of the skin all over our bodies; in its place only a very poor unnourished substitute appeared which readily rubbed raw in many places. As a result of this, the chafing of the march had already developed large raw patches in just those places where they were most troublesome. As we never took off our clothes, the peelings of hair and skin from our bodies worked down into our under-trousers and socks, and regular clearances were made from the latter.
>
>   Douglas Mawson (excerpt from *Alone on the Ice* by David Roberts).

160 kilometers from base, Mertz died. In poor physical condition, Mawson was only able to cover 8 kilometers per day, a distance that was reduced to just four by the end of January, since his energy was sapped by the need to dress and redress his many injuries. For days at a time, he was unable to make any progress because of vicious blizzards. On February 8, he found his way to base, just in time to see the expedition's ship, *Aurora*, leaving for Australia! Fortunately, a shore party had been left to wait for him, but it was too late for the ship to turn, and Mawson found himself forced to spend a second winter in Antarctica.

The Pomori

The annals of polar exploration are rich with resilient and resourceful individuals. Take the gripping tale of four Pomori hunters, who in 1743 found themselves marooned on Edgeøya Island of the Svalbard Archipelago. For 6 years, this group of hardy individuals survived everything the Arctic could throw at them: storms, chilling cold, extraordinary deprivation, confinement, and polar bears.

Their story started when the four sailed as part of a group of 14 hunters from the village of Mezen on the White Sea coast. They planned to hunt walrus in the Svalbard Archipelago. After 8 days with favorable weather, they were blown off course toward Edgeøya Island, a place where ships rarely ventured. Before long, their vessel was icebound. The situation deteriorated over the next few days, as it seemed likely their vessel would be crushed. It was decided that a four-man party would go to the island to investigate what shelter there was, since it was known that sailors had spent the winter there several years previously in a hut. The four knew they wouldn't be gone long, and they knew the hunting would be excellent, so they carried only the barest essentials.

On reaching the island, the group found the hut, where they spent the night while a storm blew outside. The next day, they made their way back to their ship to share the news with their fellow hunters. But on reaching the shore, they discovered that part of the ice pack had gone and, with it the ship, presumably carried away by the storm the previous night. The four returned to the hut and pondered the likelihood they were now trapped on the island. Permanently. They kept a watch for their ship, but after a few days, they came to the conclusion that it had foundered (the ship never returned to port, so the assumption was probably correct).

The four faced a bleak existence stuck on an island in the middle of a polar bear breeding ground with all their ammunition expended. Worse, the island was devoid of trees and shrubbery, which meant they had nothing to burn and couldn't cook. But they resolved to try to survive. After all, what else could they do? They scoured the island and found driftwood and also a plank with a long iron hook attached and some embedded nails. It wasn't much, but for these guys, it was a lifeline. Using a primitive forge, they fashioned their newfound hardware into a sharp point that they attached to a driftwood pole. The Pomori now possessed a weapon and set out to hunt polar bears.

After killing and eating their first bear, they cut the skin for clothing and used the tendons to create a string for a bow. The nails were used to manufacture rudimentary arrows. With their bow and arrows, the Pomori killed more than 250 reindeer, along with an assortment of blue and white foxes. As winter closed in, fuel economy became vital, but they couldn't allow the fire they had set to go out. Fortunately, there was no limit to the Pomori's ingenuity. They gathered slimy loam they had found during their reconnaissance of the island and fashioned a lamp. Reindeer fat was placed into the lamp and became their source of warmth during the long winter. Food continued to be a challenge. The only vegetation on the island was moss and lichen, so the men subsisted on reindeer, fox, and bear. Water was drawn from springs and made by melting ice. To prevent scurvy, the men drank reindeer blood and ate the little grass that grew on the island. One of the men wasn't too taken with the idea of drinking reindeer blood and became bedridden, eventually dying.

How did they deal with the long years of confinement and paralyzing, mind-numbing monotony? How did they cope with the chilling cold and the appalling condition of the smoke-filled hut? Who knows, but the experience of these castaways serves as an important reference point for those who are confined in a spaceship for months on end.

The Pomori were eventually rescued when, on 15 August 1749, the hunters caught sight of a Russian trading vessel on its way to Novaya Zemlya. The ship had been blown off course and had inadvertently found itself near Edgeøya. Six weeks later, the three men were finally returned home. What these hunters achieved serves as a lesson to all explorers about faith, perseverance, ingenuity, and resourcefulness. This is why many question the utility of simulating space missions, such as the Mars500 boondoggle, which placed six humans in a tin can for 520 days.

Appendix IV: Environmental and Thermal Operating System

"ISS Live!" was developed at NASA's Johnson Space Center (JSC) under NASA Contracts NNJ14RA02C and NNJ11HA14C wherein the U.S. Government retains certain rights.

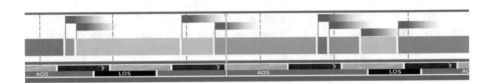

Console Handbook

ETHOS
Environmental and Thermal Operating Systems (ETHOS)

The ETHOS (pronounced *ee-those*) flight controller manages the systems which help provide a clean, safe and comfortable living area for the crew, including the monitoring of air and water onboard the International Space Station (ISS).

Every day, the ETHOS flight controller makes sure that the life support systems are working properly. This person also helps plan activities for the crew when working with environmental and thermal systems, and keeps track of the oxygen, nitrogen and water resources that are depleted, or "used up". Since it would be costly to continually deliver these resources to the ISS, the onboard life-support equipment recycles (or reuses) them as much as possible.

ETHOS
Environmental and Thermal Operating Systems

Systems Managed: ACS, ARS, ITCS, PTCS, Regen ECLSS and Emergency Response

Atmosphere Control and Supply System (ACS)
How is the cabin atmosphere created?

The comfort of the crew on the ISS is very important since they may live in space for long periods of time. It would not be practical for crewmembers to always wear pressurized suits onboard, so the Atmosphere Control and Supply System (ACS) provides oxygen and nitrogen to keep the ISS at a proper air pressure, between 14.0 and 14.9 pounds per square inch (psi). This allows the crew to live and work in an environment similar to that on Earth, and provides for proper air flow without exceeding the pressure limits of the ISS walls.

The amount of each oxygen and nitrogen in the ISS cabin atmosphere has to be monitored carefully as well. The exact mixture of these two gases is important. Some equipment may not work if the pressure is too low. If too much oxygen is present, fire can become a serious hazard on the ISS. Additionally, air pressures change when airlocks are opened for spacewalks, or when other space vehicles dock to the ISS. The ETHOS flight controller plans the appropriate amounts of oxygen (O_2) and nitrogen (N_2) required for these types of activities.

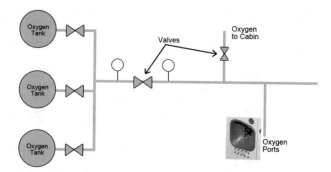

The ACS begins with oxygen and nitrogen tanks, which are attached to the outside of the ISS and have gas lines running throughout the ISS. These lines are connected to systems which use the gases (e.g., such as in onboard experiments), and to pressure control valves, which allow oxygen or nitrogen into the ISS atmosphere as needed.

Atmosphere Revitalization System (ARS)
How is fresh air maintained?

While the ACS keeps the pressure and mixture of the air ideal for healthy human living, the Atmosphere Revitalization System (ARS) monitors atmospheric gases on the ISS. Air revitalization is simply getting rid of "the bad stuff" onboard, such as carbon dioxide, air particles and contaminants.

This system consists of three main functions. The equipment (using mass spectrometry) measures the levels of gases in the atmosphere. It monitors a filter which removes air contaminants, such as chemicals released into the air by heated equipment or surfaces, crew perspiration (sweat) and fumes. It also monitors a piece of equipment which removes excess carbon dioxide (CO_2) released into the air each time the crew exhale.

The crew from Apollo 13 learned quickly the importance of atmosphere revitalization. During the mission, an onboard oxygen tank exploded. As a result, the crew was forced to shut down the fuel cells in the Service Module, which provided power, oxygen and water to the Command Module. Without power and oxygen in the Command Module, the three crewmembers had to move into the Lunar Module. In this module, the environmental systems were capable of removing the excess carbon dioxide of two people for only 30 hours, instead of three people for four days. The flight controllers in the Mission Control Center had to quickly come up with a way to make carbon dioxide filters in the Command Module fit the Lunar Module. They succeeded, and in doing so, created a working atmosphere revitalization system "on the fly".

To learn more about mass spectrometry, view the NASA video, *Mass Spectrometer: The Molecule Dissector*, at http://www.nasa.gov/multimedia/videogallery/index.html?media_id=16515038.

Internal Thermal Control System (ITCS)
How does the ISS maintain a safe and comfortable temperature?

To maintain crew safety and comfort, the ETHOS flight controller oversees the humidity (water vapor in the air) and temperature on the ISS. Too much moisture can harm the equipment and computers, causing them to fail. And while space can be very cold, the temperature onboard can vary. The ISS receives additional heat from the crew, computers and other equipment, and sunlight.

Air conditioners and fans are used to cool, circulate and remove moisture from the ISS air, which is collected and recycled into fresh water. The fans also blow air across the smoke detectors, which is extremely important in detecting a possible fire. While air conditioners and fans are helpful, the ISS utilizes an Internal Thermal Control System (ITCS), featuring cold water loop systems to help keep the computers, equipment and crew cool.

There are two internal cold water loops in the United States (U.S.) segment of the ISS, both of which fall under the responsibility of the ETHOS flight controller. The Low Temperature Loop (LTL) is kept below the dew point (the temperature at which water vapor condenses into water). It runs through the air conditioners, cooling the air and collecting condensation. The Moderate Temperature Loop (MTL) stays above the dew point and cools the computers and equipment used for experiments aboard the ISS.

In addition, pumps from the ITCS move cold water through pipes, past the warm equipment. The heat from the equipment is transferred to water in the cool pipes through conduction (a transfer of thermal energy from a substance with high temperature to a lower temperature). This keeps the crew comfortable and the equipment from overheating as they orbit the Earth. Valves in the ITCS then allow the crew to mix the cold water with warm water to keep it at the right temperature; much like faucets can control the temperature of tap water.

Passive Thermal Control System (PTCS)
Why doesn't the ISS rust?

The Passive Thermal Control System (PTCS) is used in the areas of the ISS that often get too cold. In the walls of the ISS, the temperature cannot be easily controlled by the water loops. For these areas, the ETHOS flight controller can activate the PTCS, which includes electric heaters, heat pipes and insulation. This keeps the ISS above the dew point, preventing condensation from forming on the metal which could lead to mold or corrosion (such as rust).

Regenerative Environmental Control and Life Support System (Regen ECLSS)

How are air and water recycled?

ISS crewmembers need fresh water in order to drink and to prepare their food. The ETHOS flight controller ensures there is enough water onboard. Like oxygen, water is valuable. It has to be transported from Earth to the ISS. Since the ISS is over 200 miles above the Earth, fresh water supplies are not easy to obtain. Water is heavy and it is costly to transport into space. Therefore, it is important to keep and reuse as much of it as possible.

The ETHOS flight controller monitors the equipment on the ISS which collects wastewater (i.e., condensation and urine) and recycles it into clean drinking water and oxygen. This is done through the Regenerative Environmental Control and Life Support System (Regen ECLSS) – one of the most complex systems on the ISS.

This system is made up of the Urine Processor Assembly (UPA), the Water Processor Assembly (WPA) and the Oxygen Generation System (OGS). (The OGS is comprised of both the Oxygen Generator Assembly [OGA] and the Sabatier Reduction System, or more commonly referred to as the Sabatier Reactor.)

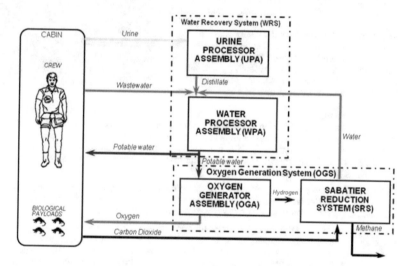

The UPA distills (heats) the collected urine to evaporate and extract the water (H_2O), which is fed into the WPA.

The WPA mixes the distilled water with the moisture collected from the air conditioners. Then, the system purifies the mixture before it is used for drinking or for making oxygen (O_2) in the OGS.

The OGS creates oxygen from water by electrolysis (the break-down of water into oxygen gas and hydrogen gas by electrical current). The OGA, the major component of the OGS, produces both oxygen (O_2) for the crew to breathe, and hydrogen (H_2) which is sent to the Sabatier Reactor. The Sabatier Reactor uses the leftover hydrogen and the excess carbon dioxide (CO_2) exhaled by the crew to create water and methane gas (CH_4).

Console Handbook – ETHOS

The water is then fed back into the system (making a complete circle of water recycling on the ISS) and the methane gas is released out of the ISS into space.

Emergency Response
How are emergencies contained on the ISS?

Crew safety on the ISS carries the most importance. Each crewmember is trained to respond to three possible types of emergencies: fire, rapid loss of pressure (when the air in the ISS leaks into space) and toxic atmosphere (ranging from a water spill to a potentially fatal ammonia spill).

Fire

If there is a fire on the ISS, unlike on Earth, the crew cannot go outside and keep a safe distance. They must put the fire out or close the hatch to the area which contains the fire. The ISS contains several smoke detectors and is designed to be non-flammable, but where there is powered equipment and oxygen, there is a risk of a fire.

If a fire is detected, an automatic response system immediately powers off all equipment and isolates any oxygen sources. All internal vents are shut down to contain contaminants and prevent fresh air from feeding the fire. If the crew can see the fire, they can use fire extinguishers similar to the ones used on Earth. If they cannot see the fire (e.g., if it is inside an experiment or a piece of equipment), they use a handheld smoke detector to help find the fire.

Rapid Depress

The ISS has a special shielding which protects the walls from being pierced by small pieces of space "junk" (broken up satellites or jettison items from EVAs) as it circles the Earth. However, if something large enough struck the ISS, it could potentially cause damage which might send the breathable air out into space. This type of emergency is called a rapid depress, which is a sudden loss of cabin pressure.

Because the air supply on the ISS is limited, the crew would need to quickly find the leak in the event of a rapid depress. Pressure sensors are located all over the ISS, and these sensors would alert the crew and the Mission Control Center of the emergency.

If a rapid depress were detected, the onboard emergency response system would automatically close all the vents which release gas overboard. The ETHOS flight controller would work with crewmembers to check each module of the ISS for leaks. By determining the location of the collision, the ETHOS flight controller and other teams in the Mission Control Center could provide instructions and a plan for the crew to repair the hole, if possible.

Toxic Atmosphere or Spill

Any spill onboard the ISS must be cleaned up immediately. Even water can damage equipment and become a breeding ground for bacteria. Without gravity, any spilled liquid or broken pieces of glass will float. This can get into the equipment or possibly injure the crew.

While water spills in the cabin can simply be wiped up, an ammonia leak can be considered fatal. The entire crew would need to be isolated from the affected areas, and would require oxygen masks to ensure that they are breathing clean air. Spills are considered an emergency on the ISS, and the ETHOS flight controller is responsible for monitoring the air quality following a spill.

To learn more about the environmental and thermal operating systems on the ISS, return to the International Space Station *Live!* (ISS*Live!*) website at www.isslive.com. Select "Interact", and then select "Visit Space Station".

ETHOS Console Displays

A wireless signal sends data from the ISS to the Mission Control Center. This data is updated on the ETHOS console displays. The current atmosphere, oxygen and water production of the ISS modules is displayed on the consoles. The ETHOS flight controller checks the data on the console displays to make sure everything is working as expected.

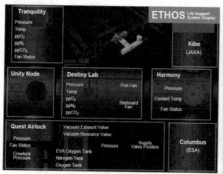
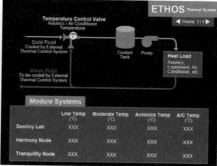

Pictured above are simplified versions of the ETHOS console displays. To view these displays, return to the ISS*Live!* website at www.isslive.com. Select "Interact", and then select "Explore Mission Control".

Space Station Live Data

To learn more about the live data streaming from the ISS to the ETHOS console display, return to the ISS*Live!* website at www.isslive.com. Select "Resources," and then select "Space Station Data". There will be a table which includes the names and brief descriptions of all the data values used to update the interactive Mission Control Center console displays.

Console Handbook – ETHOS

Acronyms and Abbreviations

ACS	Atmosphere Control and Supply System
ARS	Atmosphere Revitalization System
CH_4	methane gas
CO_2	carbon dioxide
ETHOS	Environmental and Thermal Operating Systems
H_2	hydrogen
H_2O	water
ISS	International Space Station
ITCS	Internal Thermal Control System
LTL	Low Temperature Loop
MTL	Moderate Temperature Loop
N_2	nitrogen
O_2	oxygen
OGA	Oxygen Generator Assembly
OGS	Oxygen Generation System
psi	pounds (or pounds of force) per square inch
PTCS	Passive Thermal Control System
Regen ECLSS	Regenerative Environmental Control and Life Support System
UPA	Urine Processor Assembly
U.S.	United States
WPA	Water Processor Assembly

Appendix V: EVA

EVA PREP

> **WARNING**
> Payload bay floods exceed EMU thermal limits
> during operation. If EVA crew will be operating
> in vicinity of PLB floods, floods must be turned
> off now. Cooldown time may be as long as 6 hr

MIDDECK PREP (30 min)

AW18A   1. LTG FLOOD (four) – ON
2. √EVA Bag installed in airlock
3. √REBA sw – OFF
If EMU TV:
    4. Demate EMU TV power cable; connect ground plug
5. Disconnect helmets; Velcro to lockers

HUT   6. Remove Drink Bag restraint bag
7. Fill Drink Bag from galley, remove gas and insert Drink Bag in restraint
    bag
8. Install Drink Bag restraint bag in HUT and dispose of fill tool in wet trash
9. Apply anti-fog (EMU Servicing Kit), wipe off:
    Helmets (not Fresnel lens)
    EV glasses, attach to comm cap
10. Stow EMU Servicing Kit
11. Install Helmets; lock
12. Attach Cuff C/L to EMUs

EVA PREP (90 min)

MET ____/___:___ ___	MET ____/___:___ ___	MET ____/___:___ ___

EVA PREP

PREP FOR DONNING (30 min)
If internal airlock:
ML31C   ☐☐☐   1. √VAC VENT ISOL VLV CNTL tb – OP
                  √NOZ HTR – ON

If external airlock:
BOTH DCM   ☐☐☐   2. Retrieve, position SCU; remove DCM cover
           ☐☐☐   3. Connect SCU to DCM, √locked
AW82B   ☐☐☐   4. EV-1,EV-2 O2 vlv (two) – op
MO13Q   ☐☐☐   5. √ARLK H2O S/O VLV – OP (tb-OP)
MD(flr)   ☐☐☐   6. √EMU O2 ISOL VLV – OP
ML86B:C ☐☐☐   7. √cb MNC EXT ARLK HTR ZN 1,2 (two) – op
L2   ☐☐☐   8. √O2 XOVR SYS 1,2 (two) – OP
BOTH DCM   ☐☐☐   9. PWR – BATT

> **CAUTION**
> EMU must be on BATT pwr when
> airlock power supply turned on

AW18H ☐☐☐ 10. PWR/BATT CHGR EMU 1,2 MODE (two)   – PWR
                                       BUS SEL (two) – MNA(MNB)

DCM   ☐☐☐ 11. PWR – SCU
      ☐☐☐ 12. Verify panels as shown next page

121

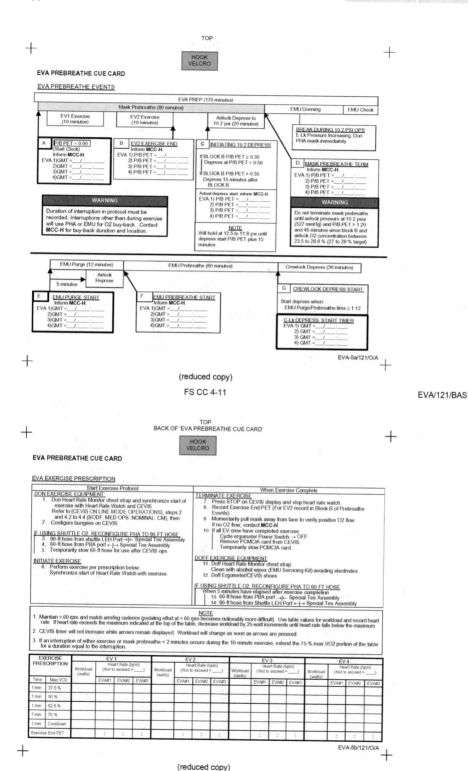

TOP

HOOK
VELCRO

EVA PREBREATHE CUE CARD

EVA PREBREATHE EVENTS

EVA PREP (170 minutes)

Mask Prebreathe (80 minutes)

EMU Donning | EMU Check

EV1 Exercise (10 minutes) | EV2 Exercise (10 minutes) | Airlock Depress to 10.2 psi (20 minutes)

BREAK DURING 10.2 PSI OPS
E-Lk Pressure Increasing. Don PHA mask immediately.

A P/B PET = 0:00 (Start Clock)
Inform **MCC-H.**
EVA 1)GMT = ___/___ ___
2)GMT = ___/___ ___
3)GMT = ___/___ ___
4)GMT = ___/___ ___

B EV2 EXERCISE END
Inform **MCC-H.**
EVA 1) P/B PET = ___/___ ___
2) P/B PET = ___/___ ___
3) P/B PET = ___/___ ___
4) P/B PET = ___/___ ___

C INITIATING 10.2 DEPRESS

If BLOCK B P/B PET ≤ 0:35
Depress at P/B PET = 0:50

If BLOCK B P/B PET > 0:35
Depress 15 minutes after BLOCK B.

Actual depress start, inform MCC-H.
EVA 1) P/B PET = ___/___ ___
2) P/B PET = ___/___ ___
3) P/B PET = ___/___ ___
4) P/B PET = ___/___ ___

NOTE
Will hold at 12.5 to 11.8 psi until depress start P/B PET plus 15 minutes

D MASK PREBREATHE TERM
Inform **MCC-H.**
EVA 1) P/B PET = ___/___ ___
2) P/B PET = ___/___ ___
3) P/B PET = ___/___ ___
4) P/B PET = ___/___ ___

WARNING

Do not terminate mask prebreathe until airlock pressure at 10.2 psia (527 mmHg) and P/B PET > 1:20 and 45 minutes since block B and airlock O2 concentration between 23.5 to 28.8 % (27 to 28 % target).

WARNING

Duration of interruption in protocol must be recorded. Interruptions other than during exercise will use PHA or EMU for O2 buy-back. Contact **MCC-H** for buy-back duration and location.

EMU Purge (12 minutes) | EMU Prebreathe (60 minutes) | Crewlock Depress (30 minutes)

Airlock Repress
5 minutes

E EMU PURGE START
Inform **MCC-H.**
EVA 1)GMT = ___/___ ___
2)GMT = ___/___ ___
3)GMT = ___/___ ___
4)GMT = ___/___ ___

F EMU PREBREATHE START
Inform **MCC-H.**
EVA 1)GMT = ___/___ ___
2)GMT = ___/___ ___
3)GMT = ___/___ ___
4)GMT = ___/___ ___

G CREWLOCK DEPRESS START

Start depress when:
EMU Purge/Prebreathe time ≥ 1:12

C-Lk DEPRESS, START TIMER
EVA 1) GMT = ___/___ ___
2) GMT = ___/___ ___
3) GMT = ___/___ ___
4) GMT = ___/___ ___

EVA-5a/121/O/A

(reduced copy)

FS CC 4-11

EVA/121/BAS

TOP
BACK OF 'EVA PREBREATHE CUE CARD'

HOOK
VELCRO

EVA PREBREATHE CUE CARD

EVA EXERCISE PRESCRIPTION

Start Exercise Protocol | When Exercise Complete

DON EXERCISE EQUIPMENT:
1. Don Heart Rate Monitor chest strap and synchronize start of exercise with Heart Rate Watch and CEVIS
Refer to {CEVIS ON LINE MODE OPERATIONS}, steps 2 and 4.2 to 4.4 (SODF: MED OPS: NOMINAL: CM), then:
2. Configure bungees on CEVIS.

IF USING SHUTTLE O2, RECONFIGURE PHA TO 90 FT HOSE
3. 90-ft hose from shuttle LEH Port→|← Special Tee Assembly
4. 60-ft hose from PBA port ←|→ Special Tee Assembly
5. Temporarily stow 60-ft hose for use after CEVIS ops.

INITIATE EXERCISE
6. Perform exercise per prescription below.
Synchronize start of Heart Rate Watch with exercise.

TERMINATE EXERCISE
7. Press STOP on CEVIS display and stop heart rate watch.
8. Record Exercise End PET (For EV2 record in Block B of Prebreathe Events).
9. Momentarily pull mask away from face to verify positive O2 flow.
If no O2 flow, contact **MCC-H.**
10. If all EV crew have completed exercise
Cycle ergometer Power Switch → OFF
Remove PCMCIA card from CEVIS.
Temporarily stow PCMCIA card.

DOFF EXERCISE EQUIPMENT
11. Doff Heart Rate Monitor chest strap.
Clean with alcohol wipes (EMU Servicing Kit) avoiding electrodes.
12. Doff Ergometer/CEVIS shoes

IF USING SHUTTLE O2, RECONFIGURE PHA TO 60 FT HOSE
When 5 minutes have elapsed after exercise completion
13. 60-ft hose from PBA port →|← Special Tee Assembly
14. 90-ft hose from Shuttle LEH Port ←|→ Special Tee Assembly

NOTE
1. Maintain > 60 rpm and match arm/leg cadence (pedaling effort at < 60 rpm becomes noticeably more difficult). Use table values for workload and record heart rate. If heart rate exceeds the maximum indicated at the top of the table, decrease workload by 25-watt increments until heart rate falls below the maximum.

2. CEVIS timer will not increase while arrows remain displayed. Workload will change as soon as arrows are pressed.

3. If an interruption of either exercise or mask prebreathe < 2 minutes occurs during the 10-minute exercise, extend the 75 % max VO2 portion of the table for a duration equal to the interruption.

EXERCISE PRESCRIPTION		EV 1				EV 2				EV 3				EV 4			
		Workload (watts)	Heart Rate (bpm) (Not to exceed = ___)			Workload (watts)	Heart Rate (bpm) (Not to exceed = ___)			Workload (watts)	Heart Rate (bpm) (Not to exceed = ___)			Workload (watts)	Heart Rate (bpm) (Not to exceed = ___)		
Time	Max VO2		EVA#1	EVA#2	EVA#3		EVA#1	EVA#2	EVA#3		EVA#1	EVA#2	EVA#3		EVA#1	EVA#2	EVA#3
1 min	37.5 %																
1 min	50 %																
1 min	62.5 %																
7 min	75 %																
1 min	Cooldown																
Exercise End PET		:	:	:		:	:	:		:	:	:		:	:	:	

EVA-5b/121/O/A

(reduced copy)

FS CC 4-12

EVA/121/BAS

Appendix VI: Hypersleep Recovery Manual

Notes for crew medical officer: the Hypersleep Recovery Scale (HRS) shall be administered every 6 hours to each crewmember for the first 24 hours following exit from hypersleep. A crewmember shall be considered functional when scoring 23 following the fourth administration of the scale.

For administration guidelines, refer to JFK Coma Recovery Scale Administration and Scoring Guidelines (Johnson Rehabilitation Institute, 2004).

Hypersleep Recovery Scale[1]

Crewmember name:	Date of hypersleep entry:			
Date of hypersleep exit:	Time of HRS administration:			
Auditory Function Scale	Exit + 6 hrs	Exit + 12 hours	Exit + 18 hours	Exit + 24 hours
4 – Consistent movement to command				
3 – Reproducible movement to command				
2 – Localization to sound				
1 – Auditory startle				
0 – None				
Visual Function Scale	Exit + 6 hrs	Exit + 12 hours	Exit + 18 hours	Exit + 24 hours
5 – Object recognition				
4 – Object localization: reaching				
3 – Visual pursuit				
2 – Fixation				
1 – Visual startle				
0 – None				
Motor Function Scale	Exit + 6 hrs	Exit + 12 hours	Exit + 18 hours	Exit + 24 hours
6 – Functional object use				
5 – Automatic motor response				
4 – Object manipulation				

	Exit + 6 hrs	Exit + 12 hours	Exit + 18 hours	Exit + 24 hours
3 – Localization to noxious stimulation				
2 – Flexion withdrawal				
1 – Abnormal posturing				
0 – None				
Oromotor/Verbal Function Scale	Exit + 6 hrs	Exit + 12 hours	Exit + 18 hours	Exit + 24 hours
3 – Intelligible verbalization				
2 – Vocalization/oral movement				
1 – Oral reflexive movement				
0 – None				
Communication Scale	Exit + 6 hrs	Exit + 12 hours	Exit + 18 hours	Exit + 24 hours
2 – Functional: accurate				
1 – Nonfunctional: intentional				
0 – None				
Arousal Scale	Exit + 6 hrs	Exit + 12 hours	Exit + 18 hours	Exit + 24 hours
3 – Attention				
2 – Eye opening without stimulation				
1 – Eye opening with stimulation				
0 – Unarousable				
Total score				

[1]Adapted from JFK Coma Recovery Scale

Interventions

Crewmembers who do not recover consciousness from hypersleep within 24 hours may enter a vegetative state (VS). The VS may be transitional en route to recovery or may progress to a long-standing and potentially irreversible condition – functional hypersleep disconnection syndrome (FHDS). FHDS is a condition from which the crewmember may not recover.

In the event that a crewmember is diagnosed as being in a VS, the CMO shall perform standard metabolic assessment of the brain using guidelines in the CMO handbook. If determination of metabolic activity cannot be made, the crewmember shall be re-scanned using quantified fluorodeoxyglucose in accordance with standard procedure. In the event that no metabolic function is determined, the crewmember may be diagnosed with FHDS. FHDS shall be scaled as mild, moderate, or severe, based on response to actions listed in the Hypersleep Disconnection Scale.

Hypersleep Disconnection Scale

Eye opening	Score	Verbal	Score	Motor	Score
None	1	None	1	None	1
To pain	2	Sounds	2	Extension	2
To command	3	Words	3	Flexion	3
Spontaneously	4	Disoriented	4	Withdraws from pain	4
		Oriented	5	Localizes to pain	5
				Localizes to pain and follows commands	6

FHDS rating: < 8, severe FHDS; 9–13, moderate FHDS; > 13, mild FHDS

The recovery of a crewmember diagnosed with FHDS will be determined by previous hypersleep disorders, age, and severity of the syndrome. Recovery stages of FHDS are coma, coming out of coma, amnesia, and memory recovery. Normally, full medication-assisted recovery takes 5 days.

Crewmembers recovering from FHDS may enter a state called post-traumatic amnesia (PTA). PTA is characterized by serious memory problems, confusion, and disorientation. Based on Earth-based hypersleep increments, those suffering from PTA normally recover within 4 days.

If recovery has not occurred after 8 days, the crewmember is deemed to have entered a permanent vegetative state (PVS) beyond which recovery is unlikely. In this event, the CMO shall consult with Mission Control to discuss the most appropriate course of action based on life support consumables.

Appendix VII: Nanotechnology

National Aeronautics and
Space Administration

Health, Medicine and Biotechnology

Nanosensor Array for Medical Diagnoses

A low-power, and compact nanosensor array chip

NASA has developed an innovative approach to improve the quality and convenience of medical diagnosis, and data transmission for immediate therapy. The new technology uses a network of nanochemical sensors on a silicon chip combined with a monitoring system composed of humidity, temperature, and pressure/flow sensors for real-time chemical and physical properties measurement of human breath for non-invasive and low-cost medical diagnosis. No such technology exists in the market today. Although many research activities are ongoing, NASAs technology is readily available for this application. With a detection range of parts per million (ppm) to parts per billion (ppb) this technology, called a nanosensor array chip, provides a highly-sensitive, low-power, and compact tool for in-situ and real time analysis. It changes the way and time decisions are made to help both patient and medical care provider to minimize their cost, optimize resources, reduce risk, and cut the amount of time needed for conducting a response.

BENEFITS

- Detection limit range: ppm to ppb
- Response time in seconds at 300 K
- Reproducible from sensor to sensor
- Low power: milliWatt /sensor
- Humidity effect is linear additive
- Easy integration (2-terminal I/V measurement)
- Sensor chip size is 1x1cm2 with 12 to 96 channels
- Non-invasive
- Low cost
- Fast and accurate
- Multi sensors for comprehensive measurement
- Wired or wireless data transmission over a long distance

technology solution

www.nasa.gov

THE TECHNOLOGY

Many diseases are accompanied by characteristic odors. Their recognition can provide diagnostic clues, guide the laboratory evaluation, and affect the choice of immediate therapy. The study of the chemical composition of human breath using gas chromatography mass spectrometry (GC/MS) has shown a correlation between the volatile compounds and the occurrence of certain illnesses. The presence of those specific compounds can provide an indication of physiological malfunction and support the diagnosis of diseases. This condition requires an analytical tool with very high sensitivity for its measurement. A number of volatile compounds, so called biomarkers, are found in breath samples, normally at low parts per billion (ppb) levels. For example, the acetone in the exhaled breath from human with other biomarkers can indicate Type I diabetes. Usually, the concentration of the volatile compounds in human breath is very low and the background relative humidity is high, almost 100%. NASAs invention utilizes an array of chemical sensors combined with humidity, temperature, and pressure for real-time breath measurement to correlate the chemical information in the breath with the state and functioning of different human organs. This tool provides a non-invasive method for fast and accurate diagnosis at the medical point of care or at home. The sensor chip includes multisensors for a comprehensive measurement of chemical composition, temperature, humidity, and pressure/flow rate. The sensor data collected from this chip can be wired or wirelessly transmitted to a computer terminal at the doctors desk or hospital monitoring center. The sensor chip can be connected directly or via Universal serial bus (USB) to a cell phone for data transmission over a long distance and receive an instruction from a doctors office for an immediate therapy.

Cell phone sensor chip   Sensor Chip

APPLICATIONS

The technology has several potential applications:

- ➲ Medical diagnosis
- ➲ Nanotechnology
- ➲ Health monitoring
- ➲ Homeland security
- ➲ Biomedicine
- ➲ Aerospace

PUBLICATIONS

Patent Pending

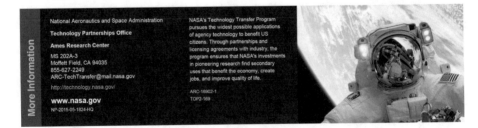

More Information

National Aeronautics and Space Administration

Technology Partnerships Office

Ames Research Center

MS 202A-3
Moffett Field, CA 94035
855-627-2249
ARC-TechTransfer@mail.nasa.gov
http://technology.nasa.gov/

www.nasa.gov
NP-2015-05-1824-HQ

NASA's Technology Transfer Program pursues the widest possible applications of agency technology to benefit US citizens. Through partnerships and licensing agreements with industry, the program ensures that NASA's investments in pioneering research find secondary uses that benefit the economy, create jobs, and improve quality of life.

ARC-16902-1
TOP2-169

Index

Printed in the United States
By Bookmasters